Electronics
in the
Evolution of
Flight

Number Nine:
Centennial of Flight Series
Roger D. Launius, General Editor

Electronics in the Evolution of Flight

Albert Helfrick

Texas A&M University Press
College Station

The paper used in this book meets the minimum requirements of the American National Standard
for Permanence of Paper for Printed Library Materials, z39.48-1984.
Binding materials have been chosen for durability.

Library of Congress Cataloging-in-Publication Data

Helfrick, Albert D.
    Electronics in the evolution of flight / Albert Helfrick. — 1st ed.
        p. cm. — (Centennial of flight series ; no. 9)
    Includes bibliographical references and index.
    ISBN 1-58544-334-4 (cloth : alk. paper) — ISBN 1-58544-413-8 (pbk. : alk. paper)
    1. Avionics. 2. Flight. 3. Aeronautics—History. I. Title. II. Series.
     TL695.H42   2004
     629.135′5—dc22

                                                  2004007214

Contents

List of Illustrations.. vi

Preface... vii

Chapter

1. The Dreamers and Philosophers 3

2. Childhood: Radio and Aeroplanes............................. 20

3. Coming of Age: The Development of Air Travel 49

4. The War Years 67

5. The Return of Air Travel 90

6. The Space Race: Lessons Learned............................ 105

7. The Computer Age: A Small Chip Starts a Big Revolution...................... 129

8. The Electronic Airplane: Learning to Fly without an Aircraft.................. 154

9. The Role of Electronics in the Second Century of Flight.......................... 175

Glossary 179

Index 183

Illustrations

U.S. Post Office airmail radio station, Salt Lake City, Utah   32

U.S. Mail facility, North Platte, Nebraska   33

Radio station KDEH, ca. 1925   34

U.S. Airmail radio station from mid-1920s   36

Oakland Airway Radio KCV, ca. 1928   37

AN range station as viewed from the air   39

Rotating light beacon used to mark airways   40

Aerial view of AN range station   41

Air traffic control in 1933   52

World's first radio-equipped control tower, Cleveland Municipal Airport   53

Cockpit of DC-3, 1939   59

Interstate airway communications station, ca. 1940   62

U.S. Army Air Corps AN/MPN-1 radar, 1943   85

CAA air traffic control center, ca. 1946   100

Air traffic control center, 1962   101

Flight coordinator moves shrimp boat along an aeronautical chart   102

Closeup view of map used by flight coordinator   103

Preface

In the annals of history, the twentieth century will always be remembered for phenomenal advances made in science, engineering, and industry. Standing in the wings ready to appear on the twentieth-century stage were Henry Ford's automobile, radio, the airplane, electrical appliances, and an industrial revolution never before experienced.

This is the story of two industries, electronics and aviation, both born at the dawn of the twentieth century. They grew to adolescence and at about age twenty-five were "married." Maturing together, they became the powerful aerospace industry of the mid- to late-twentieth century and subsequently propelled the United States to world prominence. This same industry also caused the world to live under the fear of nuclear annihilation during the cold war.

If asked what inventions were the most significant of the twentieth century, many people would certainly name the airplane, radio, television, computers, lasers, and radar. Then, asked to place all of these significant inventions into one machine and to name that machine, they would select the modern civilian or military aircraft or spacecraft.

Much of the early development of both industries was the result of the efforts of amateurs, working with very little or no money. Amateurs were a dedicated and inspired group. They made up for what they lacked in funding with enthusiasm. In the early days of aviation, barnstormers made a meager living performing at air shows and giving airplane rides for a few dollars. One well-known barnstormer risked his life to fly solo from New York to Paris. Even he found it difficult to find funding for his attempt because no one wanted to be responsible for underwriting a pilot's death. Of course, "Lucky Lindy" made it and consequently became an American hero.

These enthusiastic amateurs came from all walks of life. Farmers built airplanes in their barns. Mechanics built aircraft and engines in their garages. Even after several serious crashes, one such amateur airplane builder would not give up his quest to build an airplane. This farmer-turned-mechanic would make the name "Cessna" synonymous with small aircraft, and Cessna, in turn, made Wichita, Kansas—of all places—an aviation capital.

Like the barnstormers, radio amateurs entertained folks at county fairs by transmitting messages without charge to distant friends and relatives. They dabbled in broadcasting and advanced the state of the art away from its "home" in attic and basement "radio shacks." Some radio shacks were just that—a shack in the back yard—since the radio operator's family had banned the noisy, smelly radio activities inside the house.

These radio and aviation amateurs were very young people. One significant advantage of young, creative minds is that they haven't yet learned that something "can't be done," and so they attempt the "impossible." This may also account for why the two fields experienced rapid growth during the first quarter of the twentieth century. Just about that time, 1925, those working in aviation realized that radio was to play a very important role in the development of aircraft. Without navigation and communications, airplanes were merely expensive toys.

Of course, those working in electronics found ever-increasing applications in aircraft, to a point where there are now more electronics systems in a modern aircraft than anywhere else. In a modern aircraft more money is spent on electronics than on its engines, airframe, or interior appointments.

Aircraft led to rockets, missiles, and eventually space travel. Exploration of space would not be possible without sophisticated computers, communications systems, and automatic control systems. Most of these "space" systems are *not* in space but are part of a very large and complicated communications and tracking system on Earth.

Aviation and electronics merged to become the very large, influential, international "aerospace" industry. President Dwight Eisenhower remarked once that a powerful military/industrial complex won World War II. More than just the sophistication of the weapons, it was the ability to mass-produce high-quality products that was paramount. Eisenhower knew better than anyone else that the enormous industrial might and scientific progress of the United States was the hands-down winner of the war.

During the twentieth century, astronauts went to the moon; we sent robots to Mars and observed the beginnings of the universe. At first we communicated across the country, then across the oceans, then to astronauts on the moon, to probes in our solar system, and to signals from deep space. Television pictures from anywhere in the world came from orbiting satellites. It is amusing to note that after Americans set foot on the moon, it did not take long before the explorers brought along two very popular American icons: a television camera to beam pictures back to earth, and a car to drive on the lunar surface. Scientists probed the vast expanses of space and broke the atom apart. People also pointed "doomsday weapons" at each other during the cold war, and all of this took place during a single century.

The recorded history of the aviation and electronics industries emphasizes the players rather than their achievements. The numerous accounts of the successes of the electronics industry seem to describe mostly the consumer side of the industry. A number of books and motion pictures about the history of aviation also exist and focus on the likes of Howard Hughes, Charles Lindbergh, Chuck Yeager, and a whole cadre of astronauts. However, there was a group of unsung heroes not well known to the public. These were the scientists, engineers, and inventors who made it all happen. The history of electronics is actually the story of the unsung group of nerds who made it all happen. When crowds observe the shuttle launches from the Kennedy Space Center, the enormous power of the giant rocket engines overwhelms them. No one is impressed with how well the silent microchips keep everything going smoothly. Anyone can make a big fire. True genius is keeping a big fire under control and sending a huge rocket into a precise orbit.

The history of modern technology is chronicled in technical publications, but these are foreign territory to most people. This text outlines the important technical and engineering accomplishments in terms the moderately scientific reader can understand. We cannot appreciate the achievements of inventors and scientists if we cannot understand their scientific breakthroughs.

Abbreviations and Acronyms

Most style guides and editors disapprove of the use of abbreviations in formal writing. In the aviation world (and, more specifically, the aviation electronics world) it would be difficult to relate the true story of this industry without the use of abbreviations and acronyms. As an example, aviation electronics is called "avionics," with the "avi" from "aviation" and the "onics" from "electronics." It was not too long ago that dictionaries and spell checkers would not recognize the word, but the latest versions have been updated to include "avionics." Like avionics, other words have been used for so long that they have become a part of the language. As an example consider the word "radar." This word is an acronym for "radio detection and ranging." "Radar" is a very common word. It would be interesting to determine how many people know its true meaning.

Engineering humor created some of the abbreviations and acronyms that are now a part of the avionics story. Nowhere is this truer than in radar terms such as "fruit," "squitter," "clutter," "grass," "rabbit tracks," and "garble." The origin of some of these terms is not clear, but one thing is certain: They are not a product of a staid committee of learned lexicographers. Radar was developed with great intensity during World War II. Advancements often came about after long sleepless nights. It is not hard to imagine weary engineers and scientists having a few light moments in the early morning hours, making up a funny name for something that is very serious.

Some abbreviations are pronounced as a word if possible, while some abbreviations are never pronounced even though they can be. As an example, the Federal Aviation Administration (FAA) is never called "fa," but the FAA's parent, the Department of Transportation (DOT) is often called "dot." Some abbreviations require a bit of imagination to be pronounced, such as the "air traffic control radar beacon system" (ATCRBS). This acronym is pronounced "at-crabs." This example is also interesting as it is an acronym containing another acronym since the "R" in "ATCRBS" stands for "radar." Another example is high-frequency direction finding, or HF DF, called "huff duff."

The avionics industry has more than its share of colorful characters and funny words, and in order for the reader to appreciate the whole story, both the characters and their language need to be introduced.

In this book the more popular abbreviations are used, and when they are pronounced, the pronunciation is given. Once an acronym or abbreviation is defined, it will be used again in the book. For the reader who might have forgotten the definition, a glossary is provided.

Because the electronics world uses metric dimensions, "meters" and "kilograms" are therefore used in this text. The aviation world uses a strange mixture of measures that includes both English and metric dimensions. Fortunately, the electrical world has never used any English units. Electronics is a relatively new science, and when the tools necessary to work with electricity were invented, the scientific world realized the horrors of the English system and adopted the metric system exclusively. As an example, in this text, whenever "wavelength" is specified, it is always in meters, never feet. "Nautical miles" is the standard measure of distance for navigation and is used throughout the book. A nautical mile, abbreviated "nmi," simplified navigational calculations and was keyed to the circumference of the earth. When modern science determined Earth's circumference more accurately, the length of the nautical mile changed; it was finally defined as exactly 1,852 meters. This makes the nautical mile a "metric" measure. For altitude the unit of measurement is feet. There is no good explanation for this; altitude has always been in feet and is an international standard.

Electronics in the Evolution of Flight

Chapter 1

The Dreamers and Philosophers

Throughout the reign of the human animal on Earth, it has been human nature to be dissatisfied with one's lot in life. It is not a matter of being unhappy with one's life but rather yearning to do more. All human beings are taught that improving one's self is a necessary part of living. To most people, satisfaction comes from increasing their knowledge of the world around them or improving their skills. To others, doing what no one has done before is a necessary life enhancement. To some this means traveling over the ground faster than the cheetah. To others it means taking to the waters and sailing around the world. To still others it means swimming with the fish under the sea. One of the most intriguing desires of humankind is to fly with the birds.

Perhaps the most famous of those who dreamed of flying is Leonardo da Vinci (1452–1519), who made elaborate sketches of flying machines with which he fantasized flying like a bird. The problem with most of da Vinci's machines is that they were *too much* like birds. This drawback was not unique to da Vinci, for many other designs for a flying apparatus had wings that flapped and were called "ornithopters." One of da Vinci's sketches was for an "aerial screw" that would provide lift. When people finally took to the skies, although they had learned much from the birds, the machines they used did not resemble these creatures. Early aircraft did indeed use an airscrew, but they used it for propulsion and instead used fixed wings for lift. When aircraft construction technology improved, the early, boxy aircraft became more aerodynamic, with sweeping wings and smooth surfaces. The airscrew became the modern propeller, and at that point aircraft really began to look like birds.

Much of the dreaming of early humans was limited by their knowledge of the universe. They never dreamed of traveling to the moon and other planets because they did not know that these were places to go to. They knew of them only as points of light in the night sky. But from the moment that humankind understood the nature of the moon and the planets, it did not take very long before the dreamers wrote of traveling through space. One well-known writer, Jules Verne, wrote his first novel, *Five Weeks in a Balloon,* in 1863. Later Verne graduated to space travel with the publication of *From the Earth to the Moon* in 1865 and *Around the Moon* in 1870.

Not everyone was content with merely dreaming, and some actually took to the air. In 1783 brothers Joseph and Etienne Montgolfier successfully left the bounds of the Earth and flew in a hot-air balloon. Flying short distances produced an extreme thrill, but it was clear that a free balloon had a long way to go before becoming a practical mode of transportation. The dreamers imagined engines attached to the free balloon to ensure its path in the sky. Of course these dreamers predicted the invention of the dirigible.

Communications

Flying has always intrigued people, more because of the thrill and daring of it than its practicality. Communications, the roots of electronics, was for the most part born of necessity. Primitive tribes communicated with smoke and flags. Cave dwellers practiced written communications by drawing on the walls of their caves. With the invention of the printing press in the mid-fifteenth century, written communications got a large boost. Before the moveable-type printing press, books had to be either hand copied by scribes or printed from very expensive woodcut plates. Only the elite could afford printed materials. After the invention of the printing press, however, thousands of readers could share in religious scriptures, news, science, and the opinions of others. When electricity entered the communications arena, instant communications over long distances became a reality.

Although the thought of talking to someone a thousand miles away was intriguing, the telegraph opened up new vistas for other businesses. The railroad, for example, could not operate efficiently without the telegraph. Other businesses could expand over large geographic areas and operate as one company when tied together with the telegraph. The stock exchange could buy and sell for an entire nation with orders telegraphed to the trading floor. Some industries, such as steel, received ore from one part of the country, refined the ore with coal from another part, and shipped the product to yet another part. The dreamers envisioned huge operations spanning the entire country and delivering products to millions of consumers. In a country the size of the United States, rapid communications and efficient, well-organized transportation systems were necessary to create a single, world-class economy.

The telegraph did not reach consumers directly, in that consumers did not have a telegraph key and a sounder in their homes that they could use to communicate directly with friends and relatives. The consumer would write an outgoing message and drop it off at the local telegraph office, while a courier delivered incoming messages from the telegraph office. Telegraphy required considerable skill, and the telegraph network connected only the company's offices. However, because there were plenty of telegraph offices, the consumer could send and receive messages from the nearby "Western Union" office.

Those with vision knew that for personal communications to be really successful, a consumer should be able to communicate directly with anyone from home without the need to learn the Morse code.

The first step toward a personal communications system was taken when Alexander Graham Bell invented the telephone in 1876. But his telephone instrument was only a small part of the system that we take for granted today. If we replaced the telegraph with a telephone and used the same wires and telegraph offices, the result would be an unworkable system. In order for two customers to communicate by telephone, both parties would have to be at the telegraph (now telephone) office at the same moment. The telephone required an entirely new concept. Installing a telephone in every home required an infrastructure of wires and switching systems way beyond that used by the telegraph company. Every telephone in the system had to be capable of being connected to every other telephone, and many users had to be able to communicate without interference and with a fast connection. The telegraph companies had an infrastructure, but their "subscribers" were only the companies' own offices. If each telegraph office served a community of, say, several thousand souls, to provide each citizen with their own connection would mean the infrastructure would have to be increased by a factor of several thousand. In fact, the technologies used for the early telephone systems were quite different from those for the existing telegraph.

The revenue potential of the telephone was tremendous. Virtually everyone in the country would be a customer. Consequentially, money was spent on a system of telephone "central offices" connected with trunk lines, and the United States was wired from coast to coast. The United States was not the only nation to discover wired communications, and similar situations existed in all the developed nations of the world and many of their colonies.

Of course we all know the final result of modern electronic communications. Citizens of the industrialized world have gained access to all sorts of personal, wireless, communications devices that provide not only worldwide voice communications but also a broad spectrum of data. The modern telecommunications company is a huge worldwide conglomerate that is involved in wired, wireless, and space-based communications.

It is interesting to note that some fifty years after the invention of the telephone, the aviation industry required a coast-to-coast infrastructure of hubs (central offices), connected with airways (trunk lines) and terminating at an airport (local telephone distribution), to be successful. When the aviation infrastructure was designed, the success of the system depended on telephone, telegraph, and teleprinter communications. Where wired communication was not available or reliable, radio communication was used. Just as for the railroad industry, aviation would not be successful without long-distance communications.

Making Money with Wireless Communications

Heinrich Hertz (1857–1894) experimented with invisible waves, properly known as electromagnetic waves, which became known as "Hertzian" waves. He demonstrated several phenomena over distances of several meters, and he showed that the invisible signals could be transmitted through the air and used for wireless communications. What prevented the waves from traveling hundreds of meters? Thousands? Even millions? The reason was that the signals became weaker as the distance separating the transmitter and receiver increased. The received signal power decreased by the square of the distance. Whatever power could be received at one kilometer (km) was only one-fourth of that at two kilometers. How great a distance could be achieved by using more transmitter power or more sensitive receivers? A highly educated scientist, James Clerk Maxwell (1831–1879), had written defining equations for electromagnetic waves when Hertz was a young lad of seven. If Maxwell's equations were right, and modern science has shown they are, there are no limitations to the distances Hertzian waves can span.

Although a number of scientists were experimenting with Hertzian waves, none would become as well known as Guglielmo Marconi, for good reasons. Marconi's goal was to use Hertzian waves for two-way communications between ships at sea and distant outposts. Marconi made improved transmitters, perfected more sensitive receivers, and experimented extensively.

Marconi's vision for wireless was point-to-point communications for hire. This was more similar to the telegraph than to the telephone and would require the consumer to go to a Marconi wireless office and send what amounted to a wireless telegraph message. Since there would be no wires connecting Marconi wireless offices, the message could go directly to any Marconi facility. This could be offices worldwide, including ships at sea. Marconi exploited the safety advantages of wireless for ships and remote outposts in his quest for financial backing. Wireless did not require an expensive or expansive network with switching hubs, which was a very expensive component of the telephone system.

Even before the *Titanic*, wireless equipment was installed on ships for safety reasons. But after the *Titanic* disaster, the importance of wireless communications for ships at sea was all too evident. In the case of the *Titanic*, the wireless messages from the stricken ship went unheeded. After the *Titanic* disaster, wireless operation was taken more seriously.

During the early part of the twentieth century, the business of carrying passengers from Europe to the United States was growing. Many of the passengers were not traveling on limited budgets, so quite a bit of luxury was found on the high seas. In addition to providing safety, wireless aboard a ship was a potential gold mine. The wealthy passengers would impress their friends by sending greetings from international waters. Although Marconi is recognized as a scientist and an inventor, his major goal was to make money, and he certainly succeeded.

Getting Ready to Fly

During the last part of the nineteenth century and the very early part of the twentieth, serious experimentation in flying machines was taking place. The curious-looking ornithopters with their flapping wings were long gone, and free balloons were used only for sport. Experimenters were beginning to craft designs with fixed wings. One pioneer, Otto Lilienthal, made more than two thousand successful flights in a variety of gliders but unfortunately lost his life when one of his gliders crashed. However, he left behind valuable data on the design of wings.

During the period leading up to the advent of powered flight, scientists knew quite a bit about air but not much about the atmosphere. They knew the nature of air at the surface of the Earth, but what it was like at higher altitudes was still mostly a mystery. Scientists knew our Earth was shrouded with an atmosphere that gravity held in place. They also knew the characteristics of gasses and could easily calculate the density of air at various altitudes. The temperature of the atmosphere, particularly at higher altitudes, was not well known and actually has a lot to do with the motion of the air at these heights.

The atmosphere determines our weather, but at the end of the nineteenth century very little was known of weather fundamentals. Of course, no one knew what existed above 36,000 feet, what we today call the *stratosphere*. For early attempts to fly, this did not matter because all of the early flying experiments took place in the lower strata of the atmosphere, which people knew much more about. Surprisingly, more of our early knowledge of the atmosphere came from radio operators rather than aircraft operators. They discovered that the atmosphere played an important role not only in the flight of aircraft, but also in the propagation of radio signals. Electronics would play a role in the history of aviation once again.

Around the turn of the century, scientists believed it would be possible to generate lift and fly a heavier-than-air craft. Of course, like so many objectives, possibility is one thing, and practicability is another. On the other hand, if accurate scientific data indicate that something is impossible, then pursuing that thing is a waste of time.

At the end of the nineteenth century, electronics and aviation were at essentially the same stage. In aviation, people had flown, but scientists were convinced that the impractical, powered, heavier-than-air craft could fly. In wireless communications, Marconi had communicated with radio, but not at distances that would make it valuable, such as spanning the oceans. But Maxwell's equations predicted that it was possible. The theorists had their say, and it was now up to the pragmatists to create viable machines. Little did these early pragmatists know that their ventures would become the basis of the powerful aerospace industry of the mid-twentieth century and one day permit humans to travel to the moon.

In a historical discussion of the role of electronics in the evolution of flight, we must abandon the term *electronics* for the early stories. At the end of the nineteenth century, the word "electronics" was unknown because the role of the electron was not known. The equivalent stage in aviation would be not to know that it is air that holds up aircraft. That was never the case, however, because people can see and feel air, so there was never any question as to what held aircraft up. It is fascinating to realize that electric lights illuminated city streets and citizens were riding the streets in electrified trains and ascending to the tops of high buildings in elevators—all without a full understanding of the physics of electricity. Unlike air, the elusiveness of the electron was that it could not be seen or felt. It took nearly the first quarter of the twentieth century to fully define the electron.

In electrical engineering, a clear division of the disciplines was beginning to appear. One was electrical and the other was wireless. Eventually there would be electrical engineers and radio engineers. The electrical part included motors, lighting, heating, and the generation and distribution of the energy these devices required. Wireless, or radio, was more abstract and futuristic.

The Brothers Wright and Guglielmo Marconi

At the dawn of the twentieth century, the world was waiting for visionary inventors to implement practical applications for scientific theory. The work of Guglielmo Marconi and of Wilbur and Orville Wright was very similar in that it perfected what many others had only started. The difference is that Marconi's work applied a well-publicized scientific theory—that of Hertzian waves. The Wrights' work was based more on empirical data than scientific theory. In addition, the Wrights and Marconi fully intended to commercialize their respective fields, and they did. As the twentieth century opened, the efforts of the Wrights and Marconi would affect the world forever.

It is interesting to note that Marconi shared the Nobel Prize for his practical application of Hertzian waves. The Wright brothers, however, were never awarded a Nobel Prize despite the importance of their efforts. It is true that the Wrights did not discover any previously unknown scientific truth. They did not invent the wing, and they were not the first to fly or design a powered aircraft. Clearly, if birds could fly, so could people; thus one could argue that the Wrights only learned what the birds already knew. But Marconi did not make any fundamental scientific discoveries, either. Maxwell had already written the defining equations, and Oliver Heaviside refined those equations and wrote them in a more understandable form. Hertz had demonstrated his waves for short distances. But since no human or animal had communicated using Hertzian waves before the time of Hertz, communicating over any significant distance would be a first. Therefore, Marconi and not the Wrights received the Nobel Prize.

Wilbur Wright was born on April 16, 1867, on a farm near New Castle, Indiana. Wilbur's younger brother, Orville, was born in Dayton, Ohio, on August 19, 1871. The brothers' father was a bishop in the United Brethren Church. Neither brother earned a high school diploma. Although Wilbur had finished his prescribed courses, he decided that he did not need a diploma, so he elected not to attend graduation exercises. Orville also attended high school but decided to take only those courses he liked; unfortunately, they did not lead to a diploma.

Mechanical devices fascinated the brothers. Orville started a printing business with a press of his own manufacture. Later he published a newspaper with Wilbur as the editor. In 1892 the brothers started a bicycle sales and rental business and eventually sold bicycles of their own design and manufacture. Flying intrigued Orville and Wilbur. Off and on for about three years—from Lilienthal's death in 1896 to around 1899—the Wrights read scientific publications on aeronautics. In spite of their lack of diplomas, they quickly assimilated the available knowledge about aeronautics and in 1899 experimented with a large bi-wing kite.

After consulting the United States Weather Bureau, the Wrights selected a narrow strip of sand on the Atlantic Ocean. They were looking for an open area that provided the benefits of a prevailing wind. The site they selected was the beach called Kill Devil Hill near Kitty Hawk, North Carolina. In a letter of August 18, 1900, the weather bureau informed the brothers that the beach at Kitty Hawk generally afforded a steady wind of 10 to 20 miles per hour. That same year the Wright brothers tested an unpowered glider at Kitty Hawk. They demonstrated the ability to control the glider, and they also learned that the information provided by the weather bureau was not accurate. On the Wrights' first trip, the winds at Kitty Hawk were more than 25 miles per hour.

Encouraged by their success of 1900, the brothers returned to Kitty Hawk in 1901 with a larger glider. On this second trip the brothers from the fields of Ohio were introduced to what those who live on the shore of the North Atlantic know only too well: eighty-five-mile-per-hour winds from a winter storm. This interruption lasted for several days, after which the Wrights' experimentation continued.

Wireless and Marconi

Guglielmo Marconi was born on April 25, 1874, to a wealthy family in Bologna, Italy. He was educated at the family's estate by private tutors and later studied at the University of Bologna but, like the Wrights with their high school diplomas, never received a degree. Marconi was a quick learner and grasped complex ideas with ease. He was fascinated by Hertzian waves and, in the same spirit as the Wrights, read all he could on the subject. At the age of twenty, he set up a transmitter and receiver and transmitted wireless signals a greater distance than anyone else had done at that time.

Like the Wrights, Marconi was an astute businessman and approached the Italian government for financial backing to develop practical applications of Hertzian waves. When the government turned him down, Guglielmo traveled to England to seek the necessary funds. His change of venue paid off, and in 1897 he started the Marconi Wireless Company, Limited, with the financial assistance of the British government. About this time Marconi received the first patent—his famous number 7,777—from the English patent office for a wireless device. One of the first contracts awarded to the fledgling wireless company was for the installation of wireless stations in lighthouses along the English coast. During this time Marconi succeeded in spanning the English Channel with Hertzian waves—a distance of eighty-five miles.

During these early days of the Marconi Wireless Company, wireless stations were installed in both lighthouses and lightships. During an unusually bad storm, the Goodwin Sands lightship was seriously damaged by heavy seas. By the use of wireless, a rescue team was dispatched to the ship to prevent any loss of life. Although this rescue was from a stationary lightship, it was clear that wireless would play an important role in saving lives—first, on ships on the high seas and later in aircraft.

Marconi continually strove to cross ever-greater distances with Hertzian waves. According to Maxwell's equations, the transmitter's power, the antenna's effectiveness, and the receiver's ability to intercept faint signals were the only limits to the distances the waves could span. Marconi's goal was to cross the Atlantic Ocean with wireless signals.

Nikola Tesla, the brilliant but eccentric scientist and inventor, was convinced that Hertzian waves had no boundaries. In his February, 1901, article, "Talking with the Planets," in *Colliers Weekly* he suggested that Hertzian waves could be used to communicate with Mars. Tesla's concern was not with the ability of Hertzian waves to span the enormous distance but with the Martians' ability to understand the Morse code. Tesla, who was predisposed to showmanship, often criticized Marconi. Although Tesla was successful in working with alternating-current machines and dabbled extensively in wireless, he did not produce any useful inventions in that area. When others performed some demonstration of wireless, Tesla claimed to have thought of the same idea earlier. If Tesla had been alive when the first pictures from Mars were transmitted with Hertzian waves, he would probably have claimed that he had had the same idea nearly a hundred years ago.

After building a powerful transmitting station at Poldhu in Cornwall, England, Marconi set sail for the coast of Canada at St. Johns, Newfoundland, and arrived on December 6, 1901. He did not publicize his attempt to span the Atlantic. Although he had achieved a certain amount of notoriety for his previous experiments, where he had covered a distance of 250 miles, the North Atlantic was eight times wider, so Marconi was not certain that he would succeed. In spite of Tesla's suggestion that Hertzian waves could reach Mars, Marconi was not sure they could even span the Atlantic.

Marconi also faced another problem. If Maxwell's equations were correct, radio waves traveled in straight lines. To span the Atlantic, however, the waves would have to follow the Earth's curved surface. But Marconi had already spanned nearly 250 miles over the curvature of the Earth. If the experiment worked, he would have no explanation for his success. Another three decades would pass before scientists would understand the ionosphere and thus be able to explain the secrets of long-distance, wireless communication.

With the help of his two assistants, Marconi assembled the wireless equipment and made ready for the experiments. A large, nine-foot kite was to support the "aerial" (antenna). After the wire antenna was attached, the kite was launched, and the stiff Atlantic Ocean winds, the same winds that bedeviled the Wrights, tore the kite from the wire and sent it out to sea. Expecting trouble more from too little wind than too much, Marconi brought a balloon, which he inflated with hydrogen. After reaching the limits of the antenna wire, the balloon broke away and was lost at sea. The following day, December 12, 1901, Marconi and his two assistants launched yet another kite. The blustery winds made it difficult for the three to control the kite, but they successfully launched it and raised the receiving antenna to a height of four hundred feet.

Marconi cabled his transmitting station in Poldhu and requested the transmission of the three dots, the Morse letter *S*. Sitting in an old barracks building on a bluff overlooking the North Atlantic on a cold December day in 1901, Marconi strained to hear the faint clicking in the headphones. After a while, Marconi—a bit weary from the trials of the previous days—gave the headset to his assistant, the only other person in the barracks. "See what you can hear, Mr. Kemp." Kemp donned the headphones and after a few minutes said, "Yes, there it is!" The distinct clicking of the three dots! The historic moment was shared by only two—Marconi and Kemp. Hertzian waves had spanned the Atlantic.

The Wrights Prepare to Fly

The Wrights arrived at Kitty Hawk in September, 1903, on their fourth trip, but this time they brought along a powered aircraft. After a few months of tackling problems with the aircraft, they were ready for a powered flight. The Wrights had spent the past year working out problems associated with their glider experiments. They had generated new lift data to replace those published by the aeronautical journals of the day. They had thoroughly analyzed the calculations for their aircraft and were absolutely sure it would fly. They were so sure that they tossed a coin to see which of the two brothers would be the first person to take off and land in a controlled, heavier-than-air craft. Orville won the toss.

Exactly two years and five days after Marconi's history-making event and about nine hundred miles south on the same ocean shore, Orville Wright, lying prone, was about to be the first person to fly an airplane. Orville flew for 12 seconds over a distance of 120 feet. The brothers made four flights that day, the longest being made by Wilbur, who achieved a distance of 852 feet and a duration of 50 seconds.

There are a number of interesting parallels between the work of the Wrights and Marconi. First, all these men were young when they made their historic debuts. Wilbur was 36, oldest of the group, Orville was 32, and Marconi was 27. Because the history-making events were the result of years of preparation, these men were all in their twenties when most of the work was done. All of them had a similar goal, making the world a smaller place, by air travel or by radio waves.

The turn of the century was marked by a great deal of discovery, and much of it was done by very young men. Perhaps one of the most significant events was the publication of the special theory of relativity by a twenty-six-year-old German physicist, Albert Einstein. Although Einstein is usually associated with his theories of relativity, he won a Nobel Prize for his discovery of the photoelectric effect, an important step in our understanding of the electron. This trend of discovery would not die out. The pioneers of the newly created aviation and wireless fields, although primarily very young men, also included older men, including the Wrights and Marconi, who continued to work as they aged. There were women, too, such as the aviators Harriet Quimby, Bessica Raiche, and Blanche Stuart Scott, and we should not forget them, but for various reasons it was ambitious and courageous young men who dominated the aviation and wireless fields.

Another parallel between the work of Marconi and the Wrights was that at the time of their early experiments, none of them really understood the important role the atmosphere would play. Neither Marconi nor any other scientist of the time could explain why wireless signals followed the curvature of the Earth. Although the curved propagation of Marconi's trans-Atlantic signals had more to do with Earth's shape than its atmosphere, later experiments with shorter waves, done mostly by amateurs, revealed fantastic results. Strong signals from relatively low-powered amateur transmitters were received all around the globe but only at certain times. The behavior of Hertzian waves would lead to an improved understanding of the atmosphere. Short-wavelength wireless signals are propagated by the upper levels of the atmosphere, while the lower levels of the atmosphere contain Earth's weather, which was of prime importance to aviators. Aviation and wireless would be the impetus for most of the future atmospheric research, with the lower altitudes of interest to aviation and the higher altitudes of interest to wireless engineers.

Despite the complete success of their first powered flight in 1903, the Wrights did not make the progress that others did after that year. The quest for flight was alive and very well in Europe, and the Wrights' success was not well received there. In 1906 the first European, Santos-Dumont, repeated the Wrights' feat. But that was only a beginning. Names like Gabriel Voissin, Louis Blériot, and Frederick Handley-Page would become well known for their aviation achievements.

In the United States, the Wrights were no longer alone in their achievement. Glenn Curtiss won a prize offered by *Scientific American* magazine to fly on demand. The date chosen was July 4, 1908. Curtiss successfully flew a one-kilometer straight-line course with his aircraft, the *June Bug*. Luckily for Curtiss, the chosen path was not a closed one because the *June Bug* was barely controllable. Making a turn and returning to the starting point was not a part of the aircraft's capabilities. Curtiss, a motorcycle mechanic, built his own engines and became a name to reckon with in aircraft-engine manufacturing in future years.

Both Curtiss and the Wright brothers made a point of obtaining patent protection for their ideas. Ideally this would provide a handsome income for the use of the novel ideas that inventors developed. However, patents obtained in the very early history of a technology tend to become obsolete long before the patent expires. Unless an inventor can obtain a very broad patent on a technology that will not change, early patents tend not to be as valuable as more mature ones.

Curtiss and the Wrights became involved in patent battles beginning in 1909 even after the Wrights were warned about becoming entangled in such fights. It would have been better to allow the infringers to use the patent than attempt to profit from their illegal use by court action. What invariably happens in such situations is that the inventor is distracted from developing new patents because of the requirements of the legal action and falls behind in the technology. Moreover, at a later date the Wrights might have had an opportunity to use an important Curtiss patent. Potential legal action against Curtiss's earlier infringement is a powerful bargaining tool for the use of a newer Curtiss patent. The typical solution to such a problem is to cross-license both patents and move on. The patent battles stifled not only the progress of the Wrights and Curtiss but also of other experimenters in the United States, who feared being drawn into the fray. This was all to the advantage of European inventors, who pulled ahead in aviation technology in the period from 1903 to 1913.

As early as 1913, people were thinking of equipping aircraft with wireless. One of the stumbling blocks to achieving this was the marginal performance of the Wright aircraft and the Wrights' distraction because of the patent battles. The Wrights were believers in economical design, and their aircraft had just enough power to achieve the desired goal: lifting a pilot and a passenger. Clearly there was no excess power to allow the addition of wireless equipment.

In Europe the aviation world was moving ahead by leaps and bounds. Numerous new designs appeared and were well funded. In 1912 a French-designed Deperdussin smashed the 100-mile-per-hour barrier at 127 miles per hour. In 1913 American aircraft held only two international records while the French and some Germans held the majority. This situation persisted until the beginning of World War I and caused a serious setback in the U.S. aviation industry.

Government Regulation Calms the Chaos

The achievements of Marconi and the Wrights in the first years of the new century were harbingers of things to come. But both aircraft and wireless had a very long way to go before finding enough practical applications to make the industries profitable. The two fields attracted investors, engineers, scientists, and even a large number of amateurs. Although much work was needed, a cadre of bright, enthusiastic individuals worked diligently, and progress was rapid.

Marconi's wireless signal was capable of sending messages using Morse code. This was fine for ships at sea, where a trained radio operator was on duty. It also found application for communications between distant outposts, such as between military installations and between European nations and their colonies. Here, too, trained radio operators were important. But for wireless to become practical and achieve its full potential, voice transmission was essential.

Marconi's wireless signal had another serious problem. His transmitters generated the required high-frequency waves using mechanical means. *Frequency* refers to the number of times the direction of an electrical current changes in one second. For example, in the United States, power supplies have a frequency of 60 Hertz, abbreviated 60 Hz. The unit of measure for frequency is the Hz, which was named in honor of Heinrich Hertz.

In Marconi's transmitters, an electric motor turned a device that opened and closed a switch in such a fashion that a spark was generated. Most people have heard a click or pop in a radio when a nearby light switch is turned on or off. This demonstrates that a spark signal can radiate as a Hertzian (now called radio) wave. Today's electronics engineers call this type of signal *noise*. This noise has no value.

Marconi's spark transmitter was essentially a noise generator. The frequencies of the waves it generated were not well defined. To understand what this means, consider the modern AM broadcast radio. The radio covers a frequency range from 540 kilohertz to 1,710 kilohertz. *Kilo* is the international prefix for 1,000; thus 540 kilohertz is 540,000 Hz. The abbreviation for kilo is the lowercase *k*. Therefore, the frequency range of our example is 540 kHz to 1,710 kHz. Our AM radio has a channel every 10 kHz, which means that there are 117 channels in the frequency range we are talking about. In this example, a *channel* is a band of frequencies that may be used for communications if they do not cause interference to users of other channels.

Marconi's wireless signal was not stable, which means that its frequency would wander around and move out its channel boundary. Furthermore, his spark-generated signal occupied many channels of 10 kHz each. Only about five signals like Marconi's could be assigned to a frequency range that today accommodates 117 channels. When Marconi ran his experiments, he encountered no problems with interference because his very crude signal was the only one on the air. Later, however, he experienced interference from his own wireless stations. When wireless stations owned by other companies—sometimes in other countries—eventually went on the air, serious problems erupted.

One growing group of wireless station owners consisted of amateur radio operators, who are also called "ham" radio operators. There are several different theories about how ham radio operators got their name. One theory is that, during early experiments in broadcasting, the amateur would "ham it up" to the amusement of nearby neighbors with radio receiving sets. To call this group "amateurs" is rather misleading and would be akin to calling the Wright brothers amateurs. To a certain extent the Wright brothers and the wireless amateurs were not professionals, as they were not being paid to perform research. No one was paying to fly somewhere by airplane . . . yet. No company was paying to advertise on a wireless station . . . yet. These aviation and wireless amateurs were learning, in many cases, more than the so-called professionals were. The amateurs did not need to get approval to perform an experiment; they just did it. Until there was a clear profit base, no one was going to pay for development work in wireless or aircraft. These aviation and wireless amateurs thus became the pioneers of their fields.

The explosive growth of wireless amateurs created serious interference problems. The number of amateur wireless operators in 1917 grew to four thousand, which by twenty-first-century standards is a very small number. During this time there were no regulations whatsoever. Those who wished to communicate by radio could build or buy equipment, concoct a call sign, and simply go the air on any wavelength of their choosing. With a growing interference problem, it became clear to commercial wireless stations, amateur wireless operators, and the government that the situation had to be improved. Two measures resulted from that realization: First, the federal government began to regulate wireless transmissions. Second, everyone's efforts focused on putting the spectrum-polluting spark transmitters off the air.

In 1909 a group of seven young teenage boys in the New York City area formed the Junior Wireless Club, Limited. The seven were also members of the Junior Aero Club, which was an organization of model-aircraft enthusiasts. The boys came together because of their interest in using wireless to control their model aircraft. Of course, at this time there was no wireless of any type in full-sized aircraft. The Junior Wireless Club grew in membership, and the interest of the club members began to lean more toward wireless communication and less toward aircraft. Many of the newer members were not as young as the charter members, so in 1911 the name of the club was changed to the Radio Club of America, as it remains today.

Shortly after the formation of the wireless club, the commerce committee of the U.S. Senate held hearings on a bill, S7243, to regulate radio communication. The proposed solution to the interference problem was to deny the use of wireless to all except government and commercial interests. This bill would have essentially eliminated all amateur radio experimentation. A committee from the Junior Wireless Club went to Washington to appeal to the Senate not to pass the bill. The club's president, fourteen-year-old W. E. D "Weddy" Stokes Jr. testified before the Senate committee; the bill was subsequently defeated. In 1912 a similar bill, the Alexander Wireless Bill, appeared in the Senate commerce committee, and again the radio club took action. This bill was killed in committee. In this year the membership of the club included many of the pioneers of the radio industry, including Edwin Armstrong.

There was no question, however, that some form of regulation was required if the wireless industry was ever to be successful. Beginning in 1912, the Department of Commerce began to issue licenses with call signs that stations had to use. The transmitter power was limited to one thousand watts, and amateurs were to operate on wavelengths of two hundred meters and shorter.

The whole arrangement was a sham. Two hundred meters sometimes meant 250 or 300 meters. One thousand watts easily stretched to two thousand. There was little enforcement, except for the most flagrant violations. However, this meant that, for the first time, there were some regulations. Also, the federal government established its jurisdiction over wireless transmissions. This ensured that only one set of guidelines for wireless existed throughout the country. However, the federal government had no monitoring and measuring equipment, and even if it did, it simply did not have the expertise to use it. Because the situation was so much better than before the regulations were enacted, however, most radio operators—amateur, military, and commercial— were happy. This situation persisted until April, 1917, when the United States declared war against Germany, and all amateur operations were halted.

The war ended on November 8, 1918, but the U.S. Navy liked the wartime situation of not having to share the radio spectrum with pesky amateur radio operators. The navy saw to it that amateur radio operations did not resume after the war. It took nearly a year of heavy lobbying by the American Radio Relay League, formed in 1914, and the Radio Club of America before the government permitted amateur radio operation to resume in October, 1919. This situation was an insult to the radio amateurs of the United States. The armed forces had benefited immensely from the skills of the amateur radio operators, yet after the war these amateurs found that the privileges that had enabled them to develop equipment and skills for the war effort were taken away from them.

The radio act of February 23, 1927, created the Federal Radio Commission. This commission had twenty field offices responsible for enforcing its regulations in its twenty "radio inspection districts." The districts covered the entire continental United States, the territories of Alaska and Hawaii, and the Caribbean Islands, Virgin Islands, and Puerto Rico. The Federal Radio Commission remained in operation until 1934, when the current Federal Communications Commission was formed by the Communications Act of that year. This new agency was—and remains—responsible for all telecommunications, both wired and wireless, in the United States.

A Similar Situation in the Aviation Industry

The Wrights were joined by other experimenters who developed their own ideas for aircraft and tried to improve on the Wright brothers' aircraft. Improved control was an important goal. The Wrights' first flights were in a straight line simply because the airplane was not able to turn. Increased power for the engines was crucial for increased payload and higher altitudes. Many people worked on these problems, and they made many improvements. The new names that appeared in the annals of aviation were Glenn Curtiss, Louis Blériot, Fredrick Handley-Page, Clyde Cessna, and Walter Beech, and these pioneers were by any standard just amateurs. Eventually their names would grace large companies that were certainly not run by amateurs.

In the early 1920s one did not need a license to fly. Anyone could buy or build an airplane and fly anywhere. There was some discussion of government regulation, but unlike wireless, where the problem obviously crosses state boundaries, the regulation of aircraft could be left to the states. Of course, one could argue that aircraft could cross state boundaries as easily as wireless signals. But it was also true that automobiles and even horse-drawn wagons could easily cross state boundaries, and these means of transportation were regulated by the states. A number of states and even some municipalities had already begun to regulate aircraft, so the federal government was satisfied to take a back seat.

On May 20, 1926, President Calvin Coolidge signed the Air Commerce Act into law. This bill establishes an aeronautics branch of the Department of Commerce. The branch was not only to regulate air commerce, but also to foster its development and safety. The aeronautics branch would regulate flying, license pilots, certify aircraft designs, develop navigation aids, create an airways system, and oversee many other activities. This was an important step in the development of the aviation industry, and it came not a moment too soon, as more states and municipalities were beginning to draft aviation regulations.

By 1927 the chaos of the aviation and radio industries had been brought under control by their respective governing agencies. Government money was helping to create federal airways, and the beginnings of an air transport industry were in evidence. Radio broadcasting was becoming very successful, and a tremendous amount of money was flowing into radio design and development. The time was right for the radio and aviation industries to become inseparably linked.

Further Reading

Brady, Tim, ed. *The American Aviation Experience: A History.* Carbondale: Southern Illinois University Press, 2000.

DeSoto, Clinton. *Two Hundred Meters and Down: The History of Amateur Radio.* Newington, Conn.: American Radio Relay League, 1936, 1981, 2001.

Fisk, Fred C., and Marlin W. Todd. *The Wright Brothers from Bicycle to Biplane: An Illustrated History of the Wright Brothers.* Dayton: Toddfisk, 1990.

Harrison, James P. *Mastering the Sky: A History of Aviation from Ancient Times to the Present.* New York: Sarpedon, 1996.

Howard, Fred. *Wilbur and Orville: A Biography of the Wright Brothers.* London: Hale, 1998.

Jakab, Peter L. *Visions of a Flying Machine: The Wright Brothers and the Process of Invention.* Washington, D.C.: Smithsonian Institution Press, 1990.

Kirk, Stephen. *First in Flight: The Wright Brothers in North Carolina.* Winston-Salem: J. F. Blair, 1995.

Komons, Nick A. *Bonfires to Beacons: Federal Civil Aviation Policy under the Air Commerce Act, 1926–1938.* Washington, D.C.: Smithsonian Institution Press, 1989.

Masini, Giancarlo. *Marconi.* New York: Marsilio Publications, 1999.

Millbrook, Anna Marie. *Aviation History.* Englewood, Colo.: Jeppesen-Sanderson, 1999.

Parramore, Thomas C. *Triumph at Kitty Hawk: The Wright Brothers and Powered Flight.* Raleigh: North Carolina Department of Cultural Resources, Division of Archives and History, 1993.

Petillo, Donald M. *A History in the Making: Eighty Turbulent Years in the American Aviation Industry.* New York: McGraw-Hill, 1998.

Radio Club of America. *Seventy-Fifth Anniversary Diamond Jubilee Yearbook.* New York: Radio Club of America, 1984.

Wright, Orville. *How We Invented the Airplane: An Illustrated History.* New York: Dover, 1988; reprint of original 1920 and 1953 editions.

————. *How We Made the First Flight.* Washington, D.C.: Federal Aviation Administration, Department of Transportation, Department of Public Affairs, Aviation Education Program, 1988.

Wright, Wilbur, and Orville Wright. *Miracle at Kitty Hawk: The Letters of Wilbur and Orville Wright,* ed. Fred C. Kelly. New York: Arno Press, 1951.

Chapter 2

Childhood
Radio and Aeroplanes

T he young founders of the Radio Club of America dreamed of controlling their model aircraft with wireless. The concept of using wireless in "real," not scale-model, aircraft appeared as early as 1915. This use of wireless in real aircraft does not imply a pilotless aircraft; instead, the pilot would use it for navigation and communications. The December, 1915, issue of the *Proceedings of the Radio Club of America* featured an article by L. J. Lesh titled "The Development of Radio Sets for Aeroplanes." The following month, an article titled "Portable Aeroplane and Trench Radio Sets" by William Dubilier appeared. Lesh's history is not known, but Dubilier is associated with the infamous A. Frederick Collins.

At a time when "wireless" meant only wireless telegraphy, Collins was promoting a wireless telephone. He predicted that in a short time automobiles could be equipped with Collins's wireless telephones and used to contact a nearby garage in case of mechanical breakdown. If it were easy to install a wireless telephone in an automobile, it should not be much more difficult to install a wireless telephone in an aircraft in a few years. Collins toured the United States performing demonstrations and selling stock in his company. An article by Dubilier in the August, 1908, edition of *Modern Electrics* magazine described the wireless telephone system and listed the author as an "assistant to Mr. Collins." In spite of all the demonstrations and promises, the funds received for the sale of stock were used only for further demonstrations and stock sales. No wireless telephone was ever made. In December, 1911, four officers of the Continental Wireless Telegraph and Telephone Company, the new parent of the Collins Wireless Telephone, were indicted for stock fraud and eventually convicted.

Apparently Dubilier had abandoned the Collins fraud in sufficient time to be spared an indictment, and his 1916 radio club article was not based on any of Collins's promises. Dubilier later formed a number of corporations both in the United States and Europe. His most famous company was the Dubilier Condenser Company, which grew into a large electronics-components manufacturer. In spite of Dubilier's early interest, his companies were not directly involved in the development of airborne radio equipment. However his "condensers," now called capacitors, and other components were widely used, including in airborne equipment.

The military was acutely aware of the importance of communications to the army's signal corps, a branch dedicated solely to battlefield communications. The signal corps, whose history dated back to the use of signal flags, had been around since before the appearance of wireless communications. It was the obvious agency to handle the new wireless technology. The signal corps' interest in airborne communications started during World War I but produced little in the way of useful results. Experiments were carried out shortly after the war, in 1919, using ground-based spark transmitters.

These experiments were a failure. Chapter 1 describes the problems with using spark transmitters, so the reason for their failure is apparent. Spark was used only for Morse-code transmissions. This required a radio operator who was skilled in the transmission and reception of the code. Moreover, Morse code is slow when compared with voice transmission. In addition, in 1919, experiments were one-way transmissions because, at that time, radio transmitters, particularly spark, were quite large. As a result, it was not possible to install one in an aircraft.

Communication is only one of the potential applications of wireless, however. Around 1918 the U.S. Army, in its quest to make the airplane a more formidable tool of war, experimented with radio navigation. *Radio direction finding,* or RDF, is the art of determining the direction of a source of radio transmissions. Direction finding was one of the first noncommunications applications of radio and remains in use today. To investigate its potential for aircraft use, a directional antenna consisting of large loops of wire wound between the wings of a biplane was operated with sensitive receiving equipment. An aircraft-installed direction finder used to locate ground beacons is called a *homing device.* This means that the aircraft can be guided to the ground beacon but not to other specific locations. Because the direction finder provides heading information, the aircraft can assume a heading in the direction of the ground beacon. Even while homing, the aircraft's course may not be a straight line. If the aircraft encounters a crosswind, it is constantly being blown off course. The pilot then corrects the heading of the aircraft in the direction of the beacon. The result is a curved course to the beacon via a rather unpredictable heading. However, in the early days of radio, this navigation aid was far better than anything else that was available.

If a transmitting beacon is in an aircraft rather than on the ground, and if two or more ground-based, direction-finding stations determine a fix to the aircraft, the aircraft's actual position can be determined. This technique is called *triangulation.* The bearing to the aircraft from each ground station is relayed to a common point, possibly one of the ground stations, and the aircraft's position is determined by plotting the points on a large map. The position of the aircraft is relayed to the crew via radio. This method was not desirable, however, for several reasons.

First, triangulation using ground-based receivers requires the installation of an airborne transmitter. Although this problem became less troublesome after vacuum-tube technology allowed for the production of transmitting tubes, installing transmitters aboard aircraft was much more difficult than installing receivers. Additionally, the aircraft required continuous transmission of a radio signal. In the case of military aircraft, enemy forces could use this same signal to make the same position measurement as friendly forces. Finally, the aircraft's position had to be transmitted to the aircraft. If this information were not encrypted, enemy forces could intercept it. Thus, it clearly was not a workable system. However, triangulation is still in widespread use at the time of this writing, primarily in locating aircraft that are lost and in distress, usually due to navigational system failures. Airports usually have direction-finding equipment, and the air-traffic controllers can guide lost aircraft to those airports.

An American in Paris

As discussed in the previous chapter, many of the early developments in radio resulted from the activities of radio amateurs. During World War I, however, two developments stifled much of the potential advancement. First, most amateurs were young men who entered military service. Second, the military was wary of those who could communicate around the world without supervision, so the government silenced the amateur radio operators.

On the other hand, the army understood the importance of radio communications, not only airborne but ground-based, point-to-point communications and conducted research on its own. Fortunately, the army had in its ranks most of the radio amateurs in the United States and therefore had the personnel to conduct radio research. One amateur, Edwin Howard Armstrong, the inventor of the all-important regenerative receiver, joined the U.S. Army signal corps in 1917 and was immediately commissioned a captain. Armstrong traveled to Europe to the American Expeditionary Force (AEF) in France.

The twenty-six-year-old Armstrong was assigned to the research and inspection laboratory in Paris. Numerous projects were waiting for the new captain on his arrival in Paris, including telephony between planes; the study of French, British and U.S. radio sets and the recommendation of a final design; the study of the results of French U and Y radio sets; the study of the British system; the consideration of radically new systems; direction finding on planes; and the use of receivers on planes.

Armstrong began with an evaluation of airborne communications systems. This included simple telephonic or intercom equipment to work between the pilot and an "observer." These systems were nothing more than a hands-free telephone. Another system he evaluated was a radio communications system for operation between airplanes.

Notice that Armstrong's project list specifies "telephony" between aircraft, which implies that telegraph was not acceptable. An example of a two-way radiotelephone set of the period is Western Electric's two-way telephony radio. This radio consisted of a motor generator for generating the high voltage for the vacuum tubes, an interconnection box, the radio unit, an antenna, a microphone, and two headphones. The total weight of the radio, excluding antenna, was fifty-two pounds. The antenna, according to the specification, was three hundred feet long.

Armstrong's evaluation states that the radiotelephone set was easy to tune. It had only one control for the receiver and three for the transmitter. It is difficult to imagine trying to make these adjustments while flying an open-cockpit aircraft in enemy territory.

The radio set that operated on a wavelength of about four hundred meters was tested by installing two sets in Sopwith aircraft. The maximum range for good fidelity between the Sopwiths was three miles. With the addition of a third vacuum tube in the transmitter's power amplifier, the range could be extended to five miles. However, Armstrong noted that the signal was not as clear as it was at three miles in spite of the added transmitter power. Armstrong declared the radio set a success. These early experiments with airborne radio were difficult lessons for the early engineers. The 52-pound radio was only the beginning. In addition, three hundred feet of antenna wire and a reel to store the wire (which probably added another 30 pounds) were required. The more complicated radios required a radio operator. This implied a 170-pound person with a flight suit, helmet, and parachute, which adds another 200 pounds, but this was a small price to pay for communications that could ensure the success of the airplane.

Armstrong's 1913 regenerative receiver was one of the first to take advantage of feedback in an electronic circuit. *Feedback* is the technique of taking some signal power from the output of a device or system and feeding it back to the input. *Negative* feedback means taking the difference between the input and output and amplifying it. Armstrong used the opposite type, *positive* feedback, in which the sum of the input signal and the output are fed back and amplified. Essentially, the same signal is amplified again and again, which can produce enormous amplification from a single vacuum tube. This technique provided receivers with sensitivity well beyond that of other receivers of the day.

The radio receivers used in the early wireless experiments before about 1913, including Marconi's transatlantic triumph, provided no amplification. To be successful, very high-power transmitters and huge receiving antennas were required to provide strong enough signals for the insensitive receivers.

Armstrong's receiver developments were instrumental in the introduction of wireless in aircraft.

Armstrong's second significant invention occurred while he was with the research and inspection labs in Paris. Armstrong, now a major in the signal corps, was speculating on how to use a sensitive receiver to pick up the radiation from the ignition system of an aircraft engine and then use that signal to direct antiaircraft guns. His research represents one of the first electronic warfare (EW) systems ever conceived. Later in this text we will discuss electronic warfare in greater depth.

Armstrong noticed that a nearby transmitter caused his receiver to pick up signals not in its tuning range. The nearby transmitter caused the frequency of the received signals to be converted, or *heterodyned,* to another frequency. To take advantage of this phenomenon, Armstrong developed a highly sensitive receiver and called it the "superheterodyne."

The superheterodyne receiver uses frequency-conversion techniques to change the frequency range of interest to a lower range that can more easily be amplified and processed. Since the issuance of the superheterodyne patent in 1920, the technique has been used for every type of radio receiving system. This includes broadcasting, two-way radio, television, radar, and so on. The superheterodyne receiver reigned supreme from its inception to the end of the twentieth century. Modern receiving systems of the twenty-first century are beginning to use techniques that are not true superheterodynes but still have some similarity to the 1920 patent.

The first vacuum tube capable of amplification, called the audion, invented by Lee De Forest in 1906, brought very little improvement to the radio industry until Edwin Armstrong invented applications for the device. Armstrong's feedback increased the sensitivity of radio receivers to unheard-of levels. He found that when the level of feedback in his receiver was increased further, the audion would oscillate. An oscillating amplifier, or *oscillator,* occurs when the positive feedback increases the amplification of the circuit so much that an output signal is produced without a corresponding input signal. This oscillating circuit could provide a source of clean radio waves, called "continuous wave," or CW. These clean, CW signals would occupy no more spectrum than necessary and became the replacement for spark transmitters. From 1912 to 1922 continuous-wave transmitters using vacuum tubes were the only hope of installing transmitters in aircraft.

The problem with using the audion for radio transmitters was that the vacuum tubes of the day could not produce more than a fraction of a watt of power. High power was key to radio transmission at the long wavelengths used then. However, it was only a matter of time before large transmitting tubes would be developed, as there was no secret to making high-powered transmitting tubes for the low frequencies that broadcast stations used. Making one simply involved scaling up the dimensions of the audion. This scaling, however, does not work at higher frequencies, which became a big problem in the development of later aviation electronics, particularly radar. The first broadcast station, KDKA in Pittsburgh, operated at a power level of 100 watts using the state of the art in vacuum tubes in 1920, which was nothing more than giant audions. By the middle of the first decade of broadcasting, the larger broadcast stations were operating with as much as 50,000 watts of output power.

After World War I, from about 1919 to about 1926, there were only occasional experiments with wireless in aircraft. None of these lasted for any significant period, and finally, by 1926, there was no work at all being done on airborne wireless. At this time consumer radio was in its heyday, and entrepreneurs were making fortunes in the radio business.

There were, however, some installations of radio transmitters in aircraft during this period, mostly in mail planes. These installations proved their worth, as this 1922 report from *The Book of Radio* by Charles Taussig illustrates: An airmail pilot, delayed in Cleveland by a malfunctioning radio set, loudly criticized the radio as being not only the reason for his delay but worthless. Finally, with the radio fixed, the pilot set off for Chicago. However, when the aircraft reached Chicago the ground was obscured by thick fog. The pilot circled about looking in vain for a break in the fog. After some time spent circling in the air, the pilot still had no sight of the airfield, and the fuel tank was nearly empty. Fearing for his life, the pilot used the "worthless" radio and made a general call to anyone listening. He asked them to call the Mayfair airfield and request that they launch rocket flares.

Almost immediately, flares began to pop up through the fog, and the aircraft was able to land safely. The pilot later learned that six radio amateurs had heard the distress call and called the airfield in response to his plight.

Radio Amateurs and Shortwaves

In 1917, just prior to the entrance of the United States into World War I, the Federal Radio Commission enacted new radio regulations that relegated amateur radio operators to wavelengths shorter than 200 meters. These shortwaves were considered to be worthless, and amateurs were excluded from what was considered the desirable part of the radio spectrum — that with wavelengths longer than 200 meters.

Shortwaves were uncharted territory. Amateur radio operators had little time to explore their new wavelengths before they were silenced in 1917. But now, after successfully battling the U.S. Navy, the amateurs were permitted to return to the air, and they found much to explore. Technology had improved during the war, particularly with the inventions of Armstrong, improved vacuum tubes by De Forest, and other innovations. Spark transmitters were a thing of the past, and now the amateurs built CW transmitters and sensitive superheterodyne receivers.

At first, banishment to the shortwaves seemed to be a defeat for amateur radio. Soon after the war amateur radio operators were communicating over distances of up to two thousand miles on the short, less-than-two-hundred-meter wavelengths. Furthermore, those distances were achieved with only a few hundred watts. Obviously the steady CW transmitters and sensitive receivers had something to do with these long-distance communications. However, they reckoned, there must also be something inherent in the shortwaves that was responsible for the spectacular results.

In 1921, to explore these shortwaves, the American Radio Relay League, the preeminent amateur radio organization in the United States, sent one of its most competent operators, Paul Godley of New Jersey, to Scotland with the best amateur receiving equipment available. Of course, the equipment was Armstrong's superheterodyne, which had been invented only a few years before. Godley set up a receiving station in Ardrossan, a coastal town near Glasgow. His receiving antenna was the first use of a Beverage type and was 830 feet long. At 0050 GMT on December 9, 1921, Godley heard the signals from station 1BCG in Greenwich, Connecticut, on a wavelength of 230 meters. This station had been constructed and was operated by six members of the Radio Club of America, including Armstrong. Its transmitter used vacuum tubes and operated with a power of about 250 watts. The antenna at Greenwich was a "T" type that was 70 feet high and 100 feet long. Two days later 1BCG transmitted a complete message that Godley accurately copied. Also, eight British radio amateurs correctly copied the message using much smaller antennas. In all, Godley heard thirty American stations and one Canadian station during his stay in Scotland. A year later in 1922, a similar test was performed, and 315 American amateur radio stations were heard in Europe.

The wavelength used for the tests was about 200 meters. This was the longest wavelength the new regulations permitted. Actually, the wavelength was slightly longer than 200 meters, demonstrating how the new regulations were stretched to accommodate the situation. Although the shorter wavelengths were assumed to be inferior, no scientific evidence supported such a theory. Longer waves were successful because the crude electromechanical equipment of the time could generate them. Wavelengths that were shorter than 200 meters could not be electromechanically generated.

After Godley's tests, amateurs began to wonder what waves considerably shorter than 200 meters could do, so they began to experiment with very short waves. Then in November, 1923, two amateur radio stations in the United States, 1MO and 1XAM, communicated with 8AB in France using the unheard-of short wavelength of 110 meters.

Once word of this achievement got out, many amateur radio operators experimented with wavelengths shorter than one hundred meters with even more fantastic results. Amateurs on both sides of the "pond" were communicating on a regular basis. Nevertheless, no one had any idea how it all worked. They did not know that the ionosphere, an unknown part of the atmosphere, was making all of this possible. Eventually our understanding of the atmosphere was enhanced by using radio signals to probe for information.

Commercial companies sprang up to provide "amateur" equipment, and the excitement over amateur radio grew. This so-called amateur equipment was state of the art and better than what many commercial and government stations were using. It also became apparent that the government's regulating agencies had made a serious error in giving the amateurs all of the wavelengths shorter than 200 meters. As a result, the government decided again to take away some of the amateurs' privileges. Through a series of international radio conventions, the now-useful, shortwave spectrum was divided up for government, amateur, and commercial purposes. The radio amateurs got "bands" of wavelengths at 160, 80, 40, 20, 10, and 5 meters. Interestingly, the government was not taking any chances this time. Even though there was practically no technology that would permit operation at 10 and 5 meters, the radio conventions did not give blanket approval for amateur operation on all wavelengths shorter than 10 or 5 meters. As it turned out, about a decade after the partitioning of the radio bands, amateurs were using waves as short as 5 and even 2.5 meters.

The experiments in the use of shortwaves were very important to aviation. First, communicating over long distances is important. Second, the shorter wavelength implies that efficient, shorter antennas can be used. Third, much of the success of the radio amateurs was with relatively low powers, as little as one hundred watts. It would be much easier to install a radio transmitter using 100 watts or less aboard an aircraft than the typical thousand-watt (or more) transmitter required for communications on the longer wavelengths.

Amateur Aviators

Plenty of aviation amateurs were active in this period, too. After World War I thousands of surplus aircraft were available at very attractive prices. Still, only those people with an undying love of aviation could afford even a cheap airplane. One of the lessons learned in the war was how expensive it was to own an airplane. Something was always breaking, and, unlike on a yacht or an automobile, repairs cannot simply be made when it is convenient. In addition, everything for an airplane is much more expensive, a phenomenon that even modern aviators are familiar with. A simple piece of hose for an airplane costs three times as much as the same hose for an automobile or a kitchen sink. To defray the high cost of flying an aircraft, aviators looked for high-profit activities. These included air shows, appearances in Hollywood movies, rides for hire, and the transportation of illicit liquor during prohibition.

Amateur pilots—like amateur radio operators—made their hobby affordable by building their own aircraft and performing their own maintenance. Legal, commercial aviation—particularly the carrying of passengers—was not successful in the years after World War I. Because the cost of operating an aircraft was so great, tickets prices were much more expensive than train or ocean liner tickets. Speed was another justification for the high prices: An airplane could deliver passengers to their destination in much less time than any other conveyance. That would be true if an airplane could take off and fly to the destination airport on a regular schedule, but that did not yet happen.

The same was true for carrying the mail. Air carriers could charge more per pound for mail than for passengers. However, there was no advantage to carrying mail by air if there were no significant time advantage. The U.S. Post Office was determined that mail would travel by air. The dilemma that faced the post office was that the short routes, such as Washington to New York, which was a little more than two hundred miles, provided no significant time advantage. An airmail letter mailed in the afternoon from Washington, D.C., would arrive in New York City at the same time as first-class, special-delivery mail carried by the railroad. The evening train from New York arrived at Union Station in downtown Washington, just a few blocks from the main post office, in the middle of the night. There would be more than sufficient time to sort the mail and deliver it that same day. Also, there were plenty of trains that plied the east-coast corridor from Washington, D.C., to Boston every day.

A similar situation existed for the route from Washington to Chicago. An overnight train would have a letter from Washington or New York in the Windy City by the next day. When the destination was St. Louis, Denver, or California, however, the situation changed significantly.

Flying in Bad Weather

Although an airplane was capable of making the trip from the East Coast to California in less time than a train, it seldom did. The train took three days to make the transcontinental trip. This meant *any* three days. If it was raining, the train still ran. If it was foggy, the train still ran. When it snowed, a huge plow cleared the track, and the train continued on its way. In the dark of night, the train ran all night.

The same could not be said of the airplane at that time. In order for the aviation industry to be profitable, it had to exploit the airplane's primary strength: fast travel over long distances. Until aircraft could fly reliably at night and in all types of weather, aviation would not be cost effective. All-weather flying offered obvious advantages for military aircraft as well. If an aircraft could fly in or above the clouds, it would be invisible to antiaircraft gunners and could fly more successful missions.

In 1926 the aeronautics branch of the government enlisted the aid of the National Bureau of Standards to provide not only two-way communications for aviation but also to investigate the use of Hertzian waves for navigation. Even though only seven years had passed since the last serious efforts to use wireless for aircraft (mostly the work of Armstrong), much had changed. Wireless was now "radio," and Hertzian waves were now radio waves. Spark transmitters were declared illegal and replaced with transmitters using vacuum tubes that generated continuous waves. Radio amateurs were booted out of the longer wavelengths and banished to wavelengths shorter than two hundred meters in well-defined bands of frequencies. The Federal Radio Commission served as an effective regulatory agency to keep the radio-spectrum users in their place. Broadcasting had started in 1920, and the "golden days of radio" had begun.

Another change at that time was the use of frequencies in specifying what part of the spectrum a radio station was permitted to operate in. First, wavelengths were specified in meters, which was comfortable to those who were scientifically trained. To laypersons, "meters" was a metric term and somehow foreign. To calibrate a radio dial in meters would be abhorrent to the xenophobic American consumer. To calibrate the dial in feet, on the other hand, would be unthinkable to the scientist and those who used the metric system. Frequencies, which are measured in cycles per second and are the same for both metric and English systems, were therefore used, and everyone was happy. Beginning on May 15, 1923, the Department of Commerce began assigning radio stations an operating frequency rather than an operating wavelength.

To convert from wavelength to frequency, you divide the speed of radio waves by the wavelength. Mathematically, $f = c/l$, where f is the frequency in cycles per second, c is the speed of radio waves (which is the same as the speed of light in meters per second), and l is the wavelength in meters. The speed of light, c, is 300 million meters per second. Many years later, when the field of electronics was quite mature, the cycles-per-second unit for frequency was changed to Hertz, abbreviated Hz, to honor Heinrich Hertz.

As an example, the wavelength to which the amateur radio operators were constrained was 200 meters. To convert this to frequency, use the following equation:

$$f = c/l = 300 \text{ million}/200 = 1.5 \text{ million}$$

To be a bit more scientific in this calculation, we will use the powers of ten:

$$f = c/l = 3 \times 10^8/2 \times 10^2 = 1.5 \times 10^6 = 1.5 \text{ MHz}$$

MHz is the abbreviation for megahertz. The prefix *mega-* signifies a million, or ten to the sixth power.

The Influence of Radio Broadcasting

The Westinghouse manufacturing company constructed and licensed the first broadcast station with regular programming in 1920. Amateur radio operators had experimented with broadcasting ever since the development of voice transmissions. In 1920 general amateur broadcasting was no longer legal. An amateur license did not permit one-way communications by an amateur station. Amateur radio operators could communicate only with other amateurs.

Westinghouse's station, KDKA in Pittsburgh, was a commercial venture with two fronts. Westinghouse was interested in selling radio receivers, but no radio receiver has any value if there are no broadcast stations. Westinghouse solved that problem with KDKA. On the other hand, no advertiser would buy time on a broadcast station without listeners. This situation of 1920 was like the old chicken-and-egg dilemma: Which one would come first? Westinghouse solved the dilemma by simultaneously providing both chicken and egg.

The first radio receivers for broadcast reception were of the tuned radio frequency, or TRF, type. To appreciate the importance of the timing of Armstrong's 1920 superheterodyne patent, it isn't necessary to understand the actual operation of the TRF. It is important, however, to understand that the superheterodyne was far superior to any radio receiver design thus far. The TRFs of the first years of broadcasting were acceptable, but when the manufacturers of radio receivers became licensed to use the Armstrong patent, the performance of radio receivers transcended all imagination.

Some manufacturers decided to use the superheterodyne without obtaining a license. This was the beginning of a lifetime of patent fights for Armstrong, which ended with his suicide in 1954. Unlike the Wrights, Armstrong probably would not have benefited from permitting infringers to use the patent in the hope of striking a cross-licensing deal later. The Armstrong patent was so significant that few patents were of equal value. Eventually Armstrong's widow won the patent battle.

The contribution of broadcasting to the entire radio industry was, in a word, money. Broadcasting opened up a source of funds that aircraft or even military applications couldn't even dream of. Money permitted companies to hire talented engineers, who now trained in "radio engineering." Thanks to the profits from consumer-radio receiving sets, research was carried out on components such as vacuum tubes. Improved circuits were developed. Test equipment was manufactured, and engineers established standardized test procedures and specifications. A professional organization dedicated to radio, the Institute of Radio Engineers, or IRE, was formed to exchange information and serve as the standards-issuing organization. The infusion of large amounts of cash from the golden days of radio funded the entire industry and produced new technologies for all users of radio.

Rediscovering Aircraft Radio

When the government's aeronautics branch resumed experiments with radio communications and navigation in aircraft in 1926, radio took on a whole new dimension, and advances came fast and furiously. The developments were aimed at the consumer market, for that was where the money was. This was the philosophy of David Sarnoff, the founder of the Radio Corporation of America. Marconi dreamed of radio providing point-to-point communications for hire, but only one party at a time can take advantage of a radio communications system. Sarnoff pursued broadcasting because millions of listeners could be tapped for revenue at any one time. Armstrong was a scientist and inventor, but he appreciated the fact that radio's potential wealth had worked for Sarnoff for a while. Sarnoff would be remembered not only for his exploitation of the wealth of radio and later television but also for his aggressive activities in the development of corporate power.

The popularity of radio in the consumer market encouraged manufacturers to invest heavily in new circuits and devices because of the potentially large profits. The first approach to the development of airborne radio equipment involved using the available parts of the day. However, airborne radio equipment posed serious challenges that consumer radio equipment did not. Until about 1925, radio receivers for home use were battery powered. They required a set of heavy batteries, designated A, B, and C. The weight of the receiver was perhaps twenty pounds, but the batteries weighed at least that much. Furthermore, after a few days of operation, the A battery had to be recharged. The C battery would last for years, but after a few months of operation, the B battery, not being rechargeable, had to be replaced. The weight of the batteries and need to replace and recharge them made them inappropriate for aircraft use.

The first receivers to operate from house, or alternating (AC), current began to appear about 1926. These receivers were very popular and increased the sales of radios because they required no messy batteries. This might have appeared attractive for airborne equipment, also, but airplanes at the time did not have AC power.

Radio equipment in an aircraft also had to withstand a very harsh environment. Most radio equipment would be used in an environment that was heated and protected from rain and snow. Although temperatures could range up to ninety degrees Fahrenheit in the mostly uncooled buildings of the twenties, the temperature rarely went below freezing. More importantly, the time between the ninety-degree mark and the freezing mark was usually months. There were no vibrations or shocks. Usually no one poured hydraulic fluid or gasoline on home radios either.

Life in an aircraft was another story. Many aircraft provided heat for the

This figure shows an U.S. Post Office airmail radio station in Salt Lake City, Utah. The station used a spark transmitter, located behind the typewriter *(left),* and the receiving equipment is located to the right of the operating desk. The wires attached to the wall behind the operating desk went to the antenna. This station operated only with Morse code and relayed information relative to weather and aircraft traffic only to other ground stations.

passengers, but the remote areas where the radio equipment was located were not always heated. Consider an aircraft that takes off from the desert in Phoenix, Arizona, climbs to 10,000 feet, heads to Boise, Idaho, and arrives at night. The trip could take less than one day, and the temperature at Phoenix could be one hundred degrees, whereas the nighttime temperature at 10,000 feet over Idaho could easily be below freezing.

To gain an appreciation of how difficult it was at the time to install radio equipment in an aircraft, let us consider the automobile. Aircraft are notoriously bad for vibration, much worse than automobiles. In 1926 virtually no automobiles had radios due to the engineering problems, which were primarily associated with shock and vibration. One important difference between the 1926 automobile and the 1926 airplane is that the car was quite useful without a radio; the airplane was not. If expensive radio equipment were required to make an expensive aircraft useful, so be it.

In 1926 the National Bureau of Standards, at the request of the aeronautics branch, designed a rugged receiver for aircraft applications. The re-

This is the North Platte, Nebraska, U.S. Mail facility and radio station in 1926.

ceiver used Armstrong's superheterodyne and weighed only thirty pounds. It contained a power supply that provided the equivalent of the A, B, and C batteries from the aircraft's single-battery system. Thirty pounds is not very light but is lighter than the typical home receiver of the era. This receiver, approved by the National Bureau of Standards, had no companion transmitter, for it was intended to receive navigation signals or voice transmissions. It was strictly a one-way device.

Just as important as the development of the airborne receiver was perfecting the antenna. The first broadcast receivers for home use required rather long wire antennas for broadcast reception. The simplest receiver, the crystal set, which required no source of power, could receive only local broadcast stations with a long wire antenna. The first TRF receivers used a rather long antenna but could receive more distant stations. The superheterodynes could receive stations from anywhere in the continental United States at night with a much shorter antenna. Eventually, about 1932, home broadcast receivers had built-in antennas.

An attempt at an airborne long-wire antenna was tried with a "trailing wire" antenna. This was essentially a wire that terminated in a copper cone, which was let out from the rear of the aircraft so that it would trail behind the plane. A number of characteristics made this antenna undesirable. First, the

Radio station KDEH, circa 1925. Notice the spark transmitter located behind the typewriter. By 1925 spark transmitters had become a serious problem.

antenna produced considerable drag. Second, it represented a hazard. If the antenna should break in flight, the wire would fall to the ground and possibly strike another aircraft or injure people on the ground. Finally, if the crew forgot to reel in the antenna before landing, damage to the aircraft or other aircraft could result.

A shortened antenna can never be as efficient as a full-sized one. To compensate for an inefficient receiving antenna, one can make a receiver with excess sensitivity or employ a high-power, ground transmitter. To compensate for an inefficient transmitting antenna, one can provide additional transmitter power. This is not possible in an aircraft, however, because of the severe weight penalty. Adding sensitivity to a receiver does not cost much in power or weight. Therefore, short antennas are fine for receiving but not for transmitting.

The dimensions of a full-sized antenna depend on the wavelength. The frequencies used for the first experiments with two-way aircraft radio were between 3 and 6 MHz, which represents wavelengths between 100 and 50 meters. These wavelengths are about one-tenth the length of those used for ground-to-air transmissions during World War I.

A mast antenna with a length of slightly more than a meter was developed for the two-way radio. This represents an antenna length of 1/100th to

1/50th of a wavelength for frequencies of 3 and 6 MHz respectively. An antenna for such a small fraction of a wavelength is not an efficient antenna, but reliable communications over 20 to 50 nautical miles were achieved in these early experiments.

Recall that Armstrong evaluated a pair of airborne radiotelephones during World War I and was satisfied with a range of three to five miles using a wavelength of four hundred meters and a three-hundred-foot antenna. This range was air to air, which involves inefficient antennas on both ends. The receivers Armstrong used were not his superheterodyne and therefore not highly sensitive and could not compensate for the poor antennas. The 1926 experiments with the mast involved sensitive superheterodynes and a much shorter wavelength and used air-to-ground communications. These three differences contributed to a successful experiment.

The mast still presented considerable drag but was an improvement over the trailing wire. As the maximum airspeed of aircraft increased, the mast became more of a problem. When jet aircraft first flew, the mast was not a viable solution even though a few jet aircraft actually had them. Modern aircraft still use wavelengths up to one hundred meters for communications, but modern antennas are integrated into the vertical stabilizer and thus produce no additional drag.

The airborne transmitter posed more of a problem than the receiver did. First, the transmitter required more power from the aircraft supply. Power aboard an aircraft does not come cheap. A larger generator requires a larger engine, which could require a stronger airframe, which leads to a heavier aircraft, additional consumption of fuel, and a larger fuel tank to retain the same range. It may sound like a vicious circle, and it is. Additional weight comes with a very high price tag in aircraft.

Another problem with the airborne transmitter was its higher frequencies of operation. The frequency range of 3 to 6 MHz, trivial by today's standards, was state of the art in the mid-1920s. Fortunately, the radio amateurs, now using wavelengths below two hundred meters, were building transmitters and receivers and creating a market for high-frequency components, which became the key ingredients for lightweight, airborne, shortwave transmitters.

The plan for the early implementation of airborne communications was to provide high-powered, longwave transmitters along an air route. Aircraft could receive one-way communications along the entire length of an air route. In order to report their current position, pilots must make air-to-ground communications for the part of air-traffic control called "flight following." Because of the shorter range, air-to-ground communications would be provided at critical locations such as airport terminal areas and flight service stations. These ground stations must be linked by some means of communications. If an aircraft leaves New York for Chicago, for example, to provide flight

A more up-to-date U.S. Airmail radio station from the mid-twenties. The longwave receiver is identical to the receivers used in the previous figures showing spark transmitters. However, at the center of the operating desk is a shortwave receiver and to the right is a continuous wave shortwave transmitter.

following and avoid collisions, the radio operators at each station should know the aircraft's location along the entire length of the airway. Ground stations were linked by land teletype circuits and radio links.

Marking Airways with Radio Beams

With a reasonable start toward airborne radio communications, the aircraft industry and the aeronautics branch began to consider solutions to bad-weather navigation. What was needed was a method of providing aircraft with guidance information that could penetrate the clouds. Because optical images were unable to penetrate the clouds, optical aids were therefore out of the question.

It would be helpful to understand the nature of the problem of atmospheric visibility. Our atmosphere consists of a mixture of gasses, mostly oxygen and nitrogen. It also contains a small quantity of other gasses such as carbon dioxide (CO_2) and water vapor. There are also small amounts of argon,

The Oakland, California, Airway Radio KCV, circa 1928.

ozone, and a host of gasses from pollution sources and particulates, such as dust, smoke, and pollen.

The amount of water vapor in the air varies with the weather, as everyone knows. Water vapor is a gas, but when water vapor becomes a liquid, drops forms. When the drops are large, they fall to Earth as rain. When the drops are small, they remain suspended in the air as fog. Clouds are airborne fog. Alternatively, one could say fog is clouds that have settled to the Earth's surface. The result is the same: It is difficult to see through fog or clouds.

Light penetrates fog and clouds. Obviously, even on a cloudy day you can determine night from day. You can't see the sun, but you can see the light from the sun. Aircraft flying over the clouds can be heard but not seen. Light passes through the clouds by a phenomenon called *scattering*. Light energy is reflected from each tiny water drop. The wavelength of visible light is about one-half of a millionth of a meter, or one-half of a micrometer. The drops of water in fog are very small, on the order of a fraction of a millimeter in diameter. But they are quite large when compared to the wavelength of light, which is a fraction of a micrometer. When light energy is reflected from the water drops, the reflection travels in many directions. Therefore, the light path is not a straight line but a very complex path through the cloud or fog. Because the light is scattered, images cannot be seen.

When compared to light, radio waves are from hundreds of thousands to millions of times longer. Therefore, radio waves are not reflected from the small drops of liquid water in a cloud or fog. Even the much larger raindrops are smaller than all but the shortest radio wavelengths. This is why radio waves pass through rain and fog with no significant disturbance.

Obviously, the solution for navigating in bad weather would be based on radio waves. In 1925 radio wave propagation was becoming much better understood. People knew that radio waves traveled in straight lines, as does light. This property would be used to create radio beams for navigation. In certain situations, such as propagation through the ionosphere, radio waves behaved strangely and caused problems with the desired radio beams. Methods were developed to mitigate these problems.

On July 1, 1925, the U.S. Post Office began the first, regular, night airmail service from Hadley Field, near Newark, New Jersey, to Chicago. On the same day, Cleveland opened its new one-million-dollar municipal airport, which would be the site of several "firsts" in aviation electronics history. The airway lighting allowed for night flying but not bad-weather flying. Once ground contact was lost, the lights were of no value. A system of marking the airways was needed that would function in bad weather.

Even though the U.S. Post Office would benefit from a radio navigation system, the army enlisted the aid of the National Bureau of Standards to develop a radio navigation system that would permit an aircraft to fly a specific course without help from ground controllers. The development of this system used ideas from European navigation systems, particularly the German system developed by Elektrik Lorentz to guide Zeppelins on bombing runs during World War I. The basic approach of the system was to provide a ground station with overlapping directional antenna patterns as shown in Figure 1. The antenna patterns were individually tone modulated with the letters *A* and *N*, which are inverses of each other in Morse code. The modulation was amplitude modulation, the very same type used for broadcasting.

One antenna pattern was modulated with the letter *A,* or dit dah in Morse code. If an aircraft were well within the pattern of that antenna, the output of the receiver (as heard on a pair of headphones) would be a clear Morse code *A.* The other antenna pattern was modulated with the Morse letter *N,* or dah dit. If an aircraft were well within the pattern of the "N" segment, the receiver would receive a clear letter *N.* A third identifier transmitted a unique station call sign: one for the A transmitter and another for the N transmitter. The A and N modulation was interrupted every twenty-four seconds to transmit the Morse code station identifier. Every fifteen minutes the A and N modulation was interrupted to transmit a weather broadcast by voice. Special weather broadcasts were transmitted when needed.

When the aircraft enters the area where the two patterns overlap, one letter becomes weaker while the other becomes stronger until finally the two overlap completely, resulting in a steady tone. Because the system was capable of providing four courses, it was called the four-course range, or "AN" range.

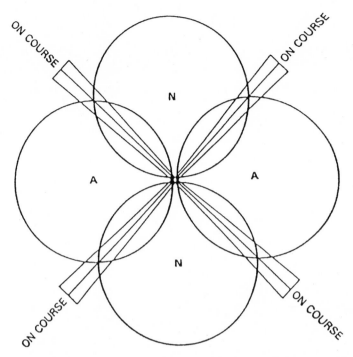

The pattern of an AN range station as viewed from the air. The antenna patterns of the four-course radio range provide an overlapping pattern for course guidance.

This system provided both navigation and broadcast communications using only one receiver and no special antenna or indicator.

The development and flight testing of the AN range was completed in February, 1928, culminating in a demonstration of the system on the New York-to-Cleveland airway. Stations were located at New Brunswick, New Jersey; Bellefonte, Pennsylvania; and Cleveland, Ohio. The first station to be turned over from the Bureau of Standards to the airways division of the aeronautics branch was the Bellefonte station, and the other two stations followed shortly thereafter. The entire New York-to-Cleveland airway using AN ranges was commissioned in November, 1928.

Because of budget constraints, more light beacons were being installed on airways than AN ranges until about 1933. In 1929 additional range stations were added to permit instrument flight as far west as Omaha. Boston and Chicago were also added to the system, and in 1930 a range station was built in Key West, which permitted instrument flight to Havana, Cuba. In 1931 additional AN range stations made instrument flight from New York to San Francisco possible. In all, ninety AN range stations covering eighteen thousand miles of airways were built by the beginning of World War II.

It must be understood what "instrument" flight meant with the AN range system. First, the AN range offered no aid for landing in instrument

A rotating light beacon installation used to mark airways. The light beacons were installed as early as the 1920s. Many light beacon locations were used for later radio navigation facilities.

conditions. In a few instances, for airports close to a range station with a runway perfectly aligned with the range's beam, the on-course beam could be used for aligning the aircraft down the runway. To provide this type of guidance, the range station would have to be at the end of the runway. As an example, at a distance of thirty nautical miles, the on-course beam from the range station is two miles wide. That lack of precision may make it difficult to land on the airport property, much less on the runway.

Second, the four courses available from a range station limited the number of airports the range could serve. The range signals could serve airports not directly under a beam, but several rather complicated procedures were required to navigate to the airport.

Other than a radio receiver, no special equipment was required to fly the AN range. The pilot simply adjusted the course of the aircraft to maintain a steady tone in the headphones. If the aircraft drifted off course, the emergence of the letter *A* or *N* would indicate how the aircraft had to be flown to correct for the error.

Flying a straight and steady AN range course was pretty simple. Things got interesting when pilots had to locate their position relative to the AN range in order to get on the correct course. Early flight manuals outlined sev-

An aerial view of an AN range station.

eral procedures to do this. All of the procedures required the pilot to fly various maneuvers while listening to the AN receiver. If the pilot heard the *A* getting stronger while the *N* was getting weaker and one of the two Morse-code identifiers was getting stronger, the pilot could determine in which quadrant around the AN range station the plane was flying.

Once it became clear which quadrant the aircraft was in, the pilot would then perform other maneuvers to get on the AN course. The pilot obtained some idea of the plane's distance from the AN range station from the width of the beams. There were marker transmitters that had a range of only a few miles, which gave an indication of absolute position. However, the markers were not homing beacons, only a verification of being on course. One indicator of absolute position was present at every range station. This was the "cone of silence." When a plane passed directly over the range station, both the *A* and *N* signals and their identifiers would fade out. If the fade was complete, the pilot knew the range station was directly under the aircraft.

Because the AN range could produce false beams, there were procedures to recognize a false beam and methods of intercepting the real one. These methods required the pilot to fly maneuvers while analyzing the received signals. This involved considerable time and a lot of air space, both of which are undesirable in instrument conditions. False beams were a particular problem in mountainous regions. When thunderstorms were in the area, the AN receiver was very susceptible to static. Imagine a pilot trying to find the AN range beam listening for *A*s, *N*s, and Morse-code identifiers all in the din of lightning static. Flying the AN range required considerable skill from the air crew. In spite of all of these problems, the network of AN ranges established from coast to coast and from north to south in the continental United States served as the major radio navigation aid for about fifteen years.

The AN range system did not provide sufficient guidance to take off from and land at an airport. The AN range airway, including en-route, two-way communications, provided guidance and flight following, but good weather was required at both the departure and destination airports. Taking off, flying en route, and landing without ground reference were called "blind flight."

The Guggenheim Fund

The Daniel Guggenheim Fund for the Promotion of Aeronautics provided financial backing for the development of a demonstration of blind fight. Guggenheim was a wealthy industrialist who had a broad spectrum of interests. In addition to funding art museums and cultural events, he also had a great interest in engineering and technology. In 1925 Guggenheim had provided New York University a grant of $500,000 for the purpose of establishing an education program in aeronautics. A year later $2.5 million was provided to establish the Guggenheim Fund, whose purpose was to promote aeronautical education and thus encourage the understanding of aeronautical science; to assist in the development of commercial aircraft and systems; and to ensure the successful use of aircraft in the nation's commerce and military operations.

The fund was to be a guiding and enabling—not a controlling—organization. The work was to be done by private industry with minimal government involvement.

The mid-to-late-1920s offered many opportunities to advance the state of aviation. The Lindbergh flight of 1927 put aviation at center stage. Lindbergh's flight was a feat of daring, not one of scientific breakthrough, but it stimulated the public's interest. One of the first endeavors taken on by the fund was the problem of all-weather flight, primarily under a low ceiling or in fog. The solution for the problem was assumed to be the dispersal of fog; the development of precise, above-ground altimetry to assist in flying through fog; the discovery of methods of locating airfields in spite of fog; and/or the use of light to penetrate fog. The fund established a full-flight laboratory for the purpose of investigating these four options in 1928 at Mitchel Field on Long Island.

Finding and landing at fog-shrouded, or "socked-in," airports had been investigated not only in the United States, but in other countries as well. In England tethered balloons were floated above the fog so that aircraft could locate a field. Obviously this system was more of a hazard than a help. In both

Britain and France, an electrified cable encircled the airport on the ground, and sensitive receiving equipment was installed in aircraft. While the British "Dingley system" got some mention in the technical press because of a modicum of success, the French system remained shrouded in obscurity. The electrified-cable solution was "almost radio," but, unlike true radio signals, had very limited range.

There were a number of attempts to eliminate fog so that pilots could use visual landing techniques. A very early program to dissipate fog used Hertzian waves and was described in the July, 1909, edition of *Technical World Magazine*. The article was written by two English authors and described the work of a French engineer, M. M. Dibbs. It is unlikely that this system worked. In 1909 the dispersal of fog was intended to aid railway trains, not aircraft.

A U.S. military program was whimsically called FIDO, for "Fog, Intense, Dispersal of." It actually extended from the 1920s into World War II. The FIDO approach was to use huge heaters to raise the temperature of the air above the dew point so that fog could no longer form. This was a very expensive proposition and did not work in all geographical areas or weather situations. Also, the technique was effective only in a limited area around the airport. Although an approach to landing can extend ten to twenty nautical miles from the touchdown point, the FIDO concept would clear only the last mile at best. The full-flight laboratory experimented with FIDO in 1929 but quickly dismissed it and all other fog-dispersal schemes as not viable and concentrated its efforts instead on a radio-based guidance system.

The foundation efforts employed two aircraft, a U.S. Navy Corsair, manufactured by Vought, and a Consolidated NY-2. The latter was the platform for the blind-flight experiments because of its extremely stable characteristics and strength. In addition to the blind-fight demonstration, the foundation-sponsored efforts included the development of improved antennas and the mitigation of the effects of ignition noise from aircraft engines.

Based on more than twenty years of experience with aircraft, from the Wrights' first flight until 1925, it was determined that three systems were needed to implement blind flight. The first was an altimeter that would be sufficiently accurate to enable pilots to land and reliably plan en-route altitudes to avoid mountains and buildings. Barometric altimetry was used in aircraft, but its accuracy was on the order of one thousand feet, which is insufficient. The altimeter must also be calibrated for local atmospheric pressure, but the voice capability of the AN range stations could be used for altimeter settings.

The second item required for blind flight was a method of providing reference to the horizon. If pilots cannot see the horizon, they cannot determine up from down. To those of us who are standing on the surface of the Earth, it is easy to tell up from down even with our eyes closed because we can feel the force of gravity on our body. When one is in an aircraft, the force that is experienced is not due to gravity alone but is a combination of the acceleration due to gravity plus any acceleration due to turning. If a pilot can see the horizon, up and down are easily determined. If an artificial horizon were provided, it would substitute for the real horizon, and the problem would be solved.

The final requirement for blind flight was a method of navigation that would direct pilots to their destination. Knowing the altitude and keeping the aircraft erect are of no value unless the aircraft can successfully complete the desired course. This is particularly true of landing. It was necessary to be able to land aircraft on the runway without seeing the runway.

Barometric altimeters had been used on aircraft since the very first days of flying. The altitude needed for landing purposes is the altitude above the ground, or above ground level (AGL). The barometric altimeter will produce an accurate indication of AGL altitude, providing the nominal barometric pressure on the ground is known. Of course, in modern times, the ground atmospheric pressure is transmitted by radio to the aircraft. In the early days of instrument flying, before all airports had two-way radio communication, pilots needed an alternate method of determining AGL altitude. One such method was to transmit acoustical signals at 950 Hz to the ground and measure the time delay from the transmission of the signal to the receipt of the echo. A horn for producing these acoustical signals required the equivalent of one hundred watts of sonic power in order to achieve a maximum range of fourteen hundred feet. There were a number of problems associated with this technique. First, 950 Hz is in the middle of the range of human hearing. This means that people on ground can clearly hear the aircraft approaching an airport "honking" a horn at 950 Hz. Second, the horn required for generating the sonic energy caused excessive drag. Because of this, the idea was abandoned, and methods of improving the existing barometric altimeters were considered. In addition, navigation radio and communications stations would broadcast the local barometric pressure.

A barometric altimeter of much greater precision than was available in the late 1920s was needed and was called the "sensitive" altimeter. Paul Kollsman of the Kollsman Instrument Company had been making altimeters in the attic of his home. He had borrowed five hundred dollars from his brother to start the enterprise. Kollsman was approached by the Guggenheim Foundation to produce an improved altimeter capable of providing one-hundred-foot resolution. Kollsman produced an instrument with two concentric rotating hands similar to a clock face. The longer pointer indicated altitude in one-hundred-foot increments while the inner hand indicated altitude in one-thousand-foot increments. This is the same format today's altimeters use. Kollsman's altimeter was a modification of an existing design rather than an entirely new approach to altimetry.

For the artificial horizon, the Guggenheim Fund turned to the master of gyros, Elmer Sperry. A sketch showed how the indicator should appear, and it was shown to Elmer Sperry Sr., who then turned the project over to his accomplished son, Elmer Jr., to work with the Guggenheim Fund. Sperry had actually retired and turned the gyroscope business over to his son, but Sperry Senior agreed to come out of retirement to take the reins of the company again so that his son could devote time to the project.

Developing a Radio Landing System

The radio system posed more of a challenge. There was no existing radio navigation system to modify or serve as a model. The AN range was in the early stages of development itself when the blind-flight effort was initiated. Furthermore, the landing guidance system was to have an indicator in the aircraft that would accurately indicate how far the aircraft was off course. In the AN system, the pilot was required to listen to the *A* and *N* characters. By careful interpretation of the signals, the pilot could get some indication of how far off course the aircraft was. Aircrews on approach and landing need to direct their entire attention to spotting the runway and communicating with air-traffic control. The complicated aerial maneuvers required to use the AN system could thus not be a part of a landing system.

Although the National Bureau of Standards was involved in a number of previous aviation radio projects, the Guggenheim Fund was deliberately searching out commercial manufacturing companies so that when the designs were done, production equipment could be fabricated. The fund selected the Aircraft Radio division of Radio Frequency Laboratories in Boonton, New Jersey. Radio Frequency Laboratories had been in the radio business since 1922 and owned a number of valuable patents for radio receivers. The company had recently purchased a sizable piece of property about three miles northwest of Boonton in nearby Boonton Township. There they built a laboratory building, an airfield, and a hangar.

The laboratory building, according to a 1929 brochure, was equipped with steam heat, was well lit, and had the most modern radio laboratory facilities. The hangar had several quarters for visiting aviators so that aircraft could be brought in for several days to be fitted with radio equipment and antennas. In spite of the steam heat, the library had a large fireplace, comfortable overstuffed furniture, and a large deer head above the mantle. It also had the latest technical journals from relatively nearby organizations such as Radio Corporation of America (New York), the Bell System (New Jersey), the Institute of Radio Engineers (New York), and the National Bureau of Standards (Washington, D.C.).

The northeast corridor with Boston to the north and Washington, D.C., to the south was the heart of electronics development at the time. Within a day's drive or a train trip from Boonton's station, one could visit pioneer companies such as the Radio Corporation of America in New York City or Camden, New Jersey; Western Electric, American Telephone and Telegraph, and Bell Laboratories, all in New York City; Weston Instruments and General Instrument in Newark, New Jersey; and countless other small companies. Edwin Armstrong had set up a business in nearby Alpine, New Jersey. Even little Boonton would become the home of Measurements, Inc.; Ferris Instruments; Ballantine Laboratories; Quan-Tech; and Boonton Radio. This proximity played a role in the development of early aircraft radio sets, which contained components from the nearby companies and were tested with test equipment from local manufacturers.

Radio Frequency Laboratories had extensive manufacturing capabilities. They could make their own sheet-metal chassis and cases. They had a coil-winding facility in which they could make their own specialized components. They were even able to manufacturer their own vacuum tubes. This was extremely important for aircraft receivers because normal vacuum tubes were not sufficiently rugged to survive in an aircraft. Heading up the aircraft radio division was Dr. Lewis Hull. Hull was no stranger to aircraft radio equipment. He received his Ph.D. in physics from Harvard and had already invented the signal generator and worked on aircraft radios.

After much engineering, the little Consolidated NY-2 aircraft was fitted with the following instruments: engine instruments, a magnetic compass; an Earth inductor compass; a turn-and-bank indicator; a directional gyro; a rate-of-climb indicator; an outside air thermometer; an artificial horizon; an airspeed indicator; an altimeter; a vibrating-reed homing indicator; and a vibrating-reed marker indicator. The last five items were new instruments developed by the fund for the demonstration flight. The last two items were radio-based indicators.

The location and nature of the indicators were given considerable attention. The pointers of some instruments were broadened so that they could be read at a glance, and the instruments were carefully placed so that the most important indications were directly in front of the pilot, who could easily scan the remaining indicators. Today these considerations are called "human factors" or "ergonomics" and are a very important part of aircraft design. Doolittle and his team knew neither of these two terms.

On September 24, 1929, the sturdy little Consolidated equipped with the necessary equipment for blind flight was rolled out of a hangar at Mitchell Field, New York. Two young army lieutenants—already experienced fliers in spite of their years—took their places in the aircraft. The rear seat in the two-place aircraft was equipped with a canopy, so the pilot in the rear had no view of the outside. The pilot in the forward cockpit was a safety pilot. The aircraft was to be flown entirely by the pilot under the canvas. The forward safety pilot was Lt. Benjamin Kelsey, and the pilot in command, actually making the epic flight, was Lt. James "Jimmy" Doolittle.

The aircraft took off, flew a fifteen-mile closed course, and landed—all without the pilot being able to see the ground. To the crowd of press and government observers, this was positive proof that airplanes could be flown in all types of weather and were viable vehicles for passenger transportation and military operations.

We have noted how young many of the aviation and scientific pioneers were. The situation for aviation electronics was no different. At the time of the epic flight, Doolittle was the patriarch at 33 years of age, Dr. Hull was 28, and Kelsey was 23. Armstrong, who invented the superheterodyne receiver used in the tests, was 22 when he filed for his first landmark patent for the regenerative receiver.

In spite of the success of the demonstration of blind flight in 1929, aviation electronic systems had a long way to go before there could be scheduled airlines or bad-weather military operations. First, the landing system Doolittle used had no vertical guidance. The aircraft was put into a constant glide path and simply flown toward the ground. The Consolidated could survive that type of abuse, but landing an air transport aircraft full of passengers with that technique would certainly not generate much repeat business.

The landing system used in the 1929 demonstration used low-frequency radio signals. These signals were susceptible to a variety of atmospheric disturbances—particularly static—during thunderstorms. An instrument landing system was eventually developed in the mid-to-late-1930s, which used very high frequencies, VHF, and eliminated most of the problems of the low-frequency waves.

A much more sophisticated landing system, developed by the National Bureau of Standards, was demonstrated in September, 1931, which included three elements. Two beams operated at a frequency of 3.105 MHz and provided horizontal guidance. This was called the *localizer,* a term that still exists today. Vertical guidance was provided by the *glide slope,* another term still in use today. The operating frequency was 90.8 MHz, which was not practical in 1931 and is one of the major reasons the system was not implemented. It is also one of the reasons the Guggenheim Foundation selected commercial companies to provide the equipment for the blind-flight demonstration. Even though the National Bureau of Standards, a government agency and not a manufacturer, produced a glide-slope system, the system could not be made commercially at reasonable prices. All the equipment used in the Guggenheim Fund demonstration was production ready. The sensitive altimeter and artificial horizon were immediately produced in large numbers. It would take more than fifteen years to design a practical instrument landing system that was available for regular use.

Further Reading

Allen, William. *The Observer's Book on Radio Navigation.* Brooklyn: Chemical Publishing Company, 1941.

Doolittle, James H. *Early Experiments in Instrument Flight.* Washington, D.C.: Smithsonian Institution Report, 1961.

———. *I Could Never Be So Lucky Again: An Autobiography.* New York: Bantam Books, 1991.

Jablonski, Edward. *Doolittle: A Biography.* Garden City, N.Y.: Doubleday, 1976.

Komons, Nick. *From Bonfires to Beacons.* Washington, D.C.: Smithsonian Institution, 1989.

Radio Club of America, ed. *The Legacies of Edwin Howard Armstrong.* New York: Radio Club of America, 1990.

Radio Frequency Laboratories. *The Design of Aircraft Receivers.* Boonton, N.J.: Radio Frequency Laboratories, 1929.

———. *The Design of Broadcast Receivers.* Boonton, N.J.: Radio Frequency Laboratories, 1929.

Redfield, Holland L. *Instrument Flying and Radio Navigation.* New York: Roland Press, 1944.

Reynolds, Quentin. *The Amazing Mr. Doolittle: A Biography of Lt. Gen. James H. Doolittle.* New York: Arno Press, 1953, 1972.

Taussig, Charles William. *The Book of Radio.* New York: D. Appleton, 1922.

Chapter 3

Coming of Age
The Development of Air Travel

Carrying passengers for hire in aircraft is a concept as old as the airplane it-self. The first serious attempts at air transport for hire were in the mid-1920s. Many of the early "air liners" were appointed more like "ocean liners" and were simply economically incapable of becoming profitable. The Russian Sikorsky Il'ya Muromets first flew in 1913 and featured lavish passenger appointments. The aircraft were so heavy and underpowered that one variant could achieve an altitude of only two thousand feet and a maximum cruise speed of forty-five knots.

In 1929 the gigantic German Dornier Do X "flying boat" was introduced. This aircraft had a wingspan of 157 feet and was powered by twelve radial engines. Although the aircraft was capable of carrying up to 160 passengers, most configurations were for fewer than 60 passengers and a crew of fourteen. The passengers traveled in royal comfort, however, with elegant furniture, wall-to-wall carpeting, fancy china, and sumptuous meals. Like the Muromets, the Dornier was slow and incapable of high altitudes, although a later version was able to reach twelve thousand feet.

The secret to airline profit at the time was to take a reasonable number of passengers to their destination in the shortest time. Unlike the luxurious aircraft, successful airliners would provide passengers with nothing more than a seat. Many of the air carriers from this early period were more interested in carrying the mail than passengers. It was obvious that mail had a higher priority when passengers were often "bumped" from a flight to carry more mail. The situation was simple: The airline company made more profits per pound by carrying mail.

The fees paid to fly the mail were so high that the U.S. Post Office was doing so at a loss. When the major airmail carriers realized they received more to carry a piece of mail than the postage affixed to the letter, they became the post office's best customer. Because the air carriers were paid by the ounce, the carriers were sending thick letters to each other. Other airmail included sand, bricks, and, in one case, a cast-iron stove.

One of the requirements of a successful, air-travel system in the United States was the existence of an air carrier large enough to fly coast to coast. Air travel made sense when the trip was long, but there were no carriers with sufficient coverage to permit a passenger to travel coast to coast without changing carriers. A passenger wishing to travel from New York to Los Angeles would have to change carriers several times. To make things worse, the air travel system was not seamless. Travelers sometimes had to change airports by ground transportation. Schedules were not coordinated, so it was not uncommon for one aircraft to arrive shortly after the connecting flight had left, requiring the passenger to spend an entire day at the intermediate airport. Of the 53 airline routes in 1929, 43 were shorter than 500 miles, and 8 were greater than 500 miles but shorter than 1,000 miles. With the lack of electronic navigation and lighted airways, short routes were the only ones that were reliable. The only route that made economic sense in 1929 was the United Air Lines route from New York to San Francisco.

In the 1920s the airways were established and lighted, but bad-weather flying was not possible, and diversions to alternate airports were common. The sensible air-transport aircraft of the late 1920s—the Ford tri-motors and Fokkers—could not carry passengers and make a profit. Passengers were carried only to win airmail routes from the post office.

Between 1925 and 1929 things changed, and aviation was taken seriously. The lighting of the airways, development of radio navigation, and Doolittle's blind-flight demonstration had piqued the interest of the investment bankers, who were fat with money from a burgeoning stock market.

The first round of airmail contracts was let in 1925 and was to expire in 1929. Many of the abuses of the airmail carriers did not go unnoticed. Expecting new, lucrative contracts in 1929, new entities appeared that were not tainted by earlier transgressions. Of course, many of these new companies were the same old actors but in a new show.

The national airway map was extended from an original 14,700 miles of airmail routes to more than twice that number. The recipients of the expanded airmail routes—and thus the ticket to a successful passenger airline—were United, American, TWA, Eastern, Northwest Airways, and Pittsburgh Aviation Industries.

After the stock-market crash of 1929, most industries faced serious setbacks, but not so with the air-carrier industry. The number of passengers carried by the airlines in 1929 was 160,000. In 1932 the figure was 474,000. From 1930 to 1932 the airlines' scheduled miles increased from 16.2 million to 34.5 million. The aircraft were not empty, either, as passenger miles increased from 84 million to 127 million. *Aviation* magazine observed in March, 1931, that "Air transport has won fame as a depression-proof industry. Of how many other industries can that be said?" The article did not respond to its own question, but the answer was "radio."

Radio and Air Travel during the Depression

In 1929 three million radios were sold, and most of those models were of the console type. The depression affected the sale of the more expensive sets, so the industry responded with less expensive table models. In 1933 seventy-four percent of all radio sales were table models. Like the airline industry, radio manufacturers not only did not suffer during the depression but in fact actually flourished. In 1930 radio sales increased to 3.8 million per year, rising to 7.6 million in 1937. The depression did not stop electronic research, either. In 1929 Vladimir Zworykin and Philo Farnsworth independently demonstrated the television.

Radio and air transport worked hand in hand in the 1930s. In the early part of the decade, radio ranges were constructed, and airways were extended. Improved airborne radio communications were provided. Ground-based telecommunications, particularly teletype circuits, were beneficial to the airlines. Without long-distance communications, the airlines could not have operated their long-distance routes. Finally, in 1938 the U.S. air-transport business logged 476 million passenger miles, but the coming war years changed everything. The thrust of the U.S. aviation industry naturally became the war, and the air-transport industry took a back seat to military transport.

The key to successful air transport was an effective air-traffic control system. Although in the 1930s, radio navigation and communications were installed along many airways—with more coming on line every day—airplane equipage was variable. In its chapter on "Airway Radio Flying and Orientation," a pilot's handbook from that decade states, "If your plane is equipped with a transmitter. . . ." Many aircraft had only navigation and communications *receivers.* Pilots filed their flight plans in advance by telephone. Then, at the appointed hour, the aircraft would taxi onto the field while listening to the local control tower on the aircraft's radio. The tower would call the aircraft, which the flight crew were expecting because of the flight plan: "Lockheed 1034, this is Cleveland tower. If you read me, wiggle your rudder." Then the aircraft would receive clearance for the flight plan and permission to take off. This was instrument flight from a controlled field. Cleveland Municipal Airport, incidentally, had the first radio-equipped, air-traffic control tower in the United States in 1930.

The tower at the departure airport sent a message via teletype to all stations, typically airports along the flight path. An aircraft without a transmitter would make low passes over the airports on the flight plan, and the airway stations would observe the aircraft with binoculars and report the time of passing on the same teletype network. When the aircraft arrived at the destination airport, the pilot in command was required to appear in person at the federal airways office and close the flight plan.

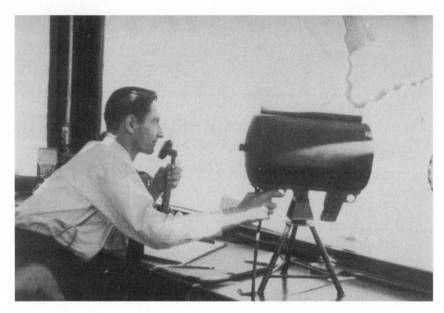

Air traffic control in 1933. The air traffic controller has his eye on the aircraft and is looking for a response from a rudder. In his left hand the air traffic controller holds a radio microphone, and just in case the aircraft does not have a radio receiver, he has his right hand firmly on a light gun.

Instrument flight in the 1930s did not include instrument landings, as the radio navigation at that time was available only en route. Larger aircraft, primarily air transport, were always equipped with transmitters and could make position reports via radio, so they did not depend on being seen at an airway station. In fact, many airlines maintained their own radio communications network. This was most important when flying in or above bad weather.

The two-way radio equipment installed aboard the airliners was used not only for communications with the federal airways but for private company communications as well. The federal airways radio stations did not pass along information about the number of passengers, needs for repairs to an aircraft, or connecting flights. Information of this nature was up to the airlines to provide.

The scheduled airlines did not wish to bear the expense of setting up their own private radio networks. Furthermore, the Federal Radio Commission did not like the idea of assigning a large number of frequencies so that each airline could have its own private radio network. There was insufficient radio traffic to justify private frequencies. On December 2, 1929, representatives from Western Air Express, Varney Airlines, American Airways, and National Air Transport met with Louis Caldwell, who had been the first general counsel for the Federal Radio Commission. The purpose of the meeting was

The world's first radio-equipped control tower at the Cleveland Municipal Airport, 1930. This tower provided an almost complete hemispherical view of the airport because of the glass roof and complete glass windows and doors. This control tower was built without tinted glass or air-conditioning. Notice the large vent on the roof and the open windows.

to create a corporation to serve as "the single licensee and coordinator of radio communication outside the government" for the transportation industry.

The corporation was named Aeronautical Radio, Incorporated. Caldwell was elected the interim president but was quickly replaced by Herbert Hoover Jr., the son of the United States president and vice-president of communications for Western Air Express. Within months, Pan American Airways, Pan American Grace Airways, and Transcontinental and Western Air Transport (the precursor to TWA) joined as stockholders of Aeronautical Radio. The charter of Aeronautical Radio stipulated, and still does today, that only airlines and aviation-related companies could own stock and that no single stockholder could own more than twenty percent. Any airline could use Aeronautical Radio's communications network and did not have to own stock in the corporation.

Immediately after the establishment of Aeronautical Radio, the Federal Radio Commission transferred all licenses previously granted to individual airlines to Aeronautical Radio. Those stations already in operation by the airlines were leased to Aeronautical Radio, who ensured that the stations were operated within the rules of the Federal Radio Commission. New stations were constructed by Aeronautical Radio as needed.

By the mid-1930s a system of airways between the major cities of the United States had been established and connected with a network of AN range stations. In addition, light- and radio-beacon stations were placed along the airways, and the major airports of the continental United States were served by airways.

Improved Radio Navigation

There were several disadvantages to the AN ranges. The first was that they provided only four courses: two inbound and two outbound. Ideally the four courses would be separated by 90 degrees. The courses did not have to be exactly 90 degrees apart, and the headings of the courses could be adjusted somewhat. However, the variation of course adjustment was restricted to some extent. Courses could not be separated by only 20 degrees, for example.

Second, the AN range was difficult to fly and required complex movements to intercept the on-course beam. The maneuvers used quite a bit of airspace, and the procedures required the aircrew to contact air-traffic control for permission before performing beam-interception maneuvers. A popular book on radio navigation at the time included 110 pages of AN range procedures for intercepting, identifying, and remaining on an AN range beam. All of the procedures required an analysis of the signals received from the range station. It was necessary to compare the strength of the A to the N or a steady tone with a faint A or N beginning to appear as the aircraft drifted away from the beam. All of this had to be accomplished in a very noisy airplane. Although it may seem convenient to use an aural indicator permitting pilots to focus their attention out the window, their aural attention should be directed to the air-traffic controller and not Morse As and Ns. A visual indicator would be more suited to the task. In fact, an indicator that gives pilots a picture of their situation would be the most appropriate.

Third, the AN range was based on long wavelengths, which are susceptible to static from thunderstorms and to another type of static—precipitation static—that the aircraft itself generates in the vicinity of thunderstorms. Although long-wavelength radio signals do not generally travel by ionospheric propagation, to a small extent they do at night. The situation is that these signals propagate by both *ground waves* and the ionosphere (in the latter case they are called *sky waves*). What happens is that the two types of signals interfere with each other and thus cause errors in the radio beams. Another problem is caused by range stations that use the same frequency but are usually out of range of each other. At night their signals propagate over much longer distances and interfere with each other. This is called the *night effect*.

Some improvements were made by changing the antenna arrays at the transmitting stations. The stations began to employ multiple towers that were interconnected with specific lengths of transmission line. In this way, the radio signals were transmitted at shallower take-off angles, and the energy directed to the ionosphere was thus reduced. This transmission-line (TL) version of the AN range station improved but did not completely eliminate the night effect.

Finally, the low-frequency waves used by the AN ranges propagated by ground wave, which was very strongly affected by the nature of the soil, particularly its conductivity, around the transmitting site. A sudden change of conductivity such as from land to water, particularly the more conductive salt water, would cause the beams to bend. This was called the *shore effect.*

In mountainous areas, the rock outcroppings are less conductive than the valleys they form. This causes false beams, which are even more deleterious than bent beams. If a beam has a bend, and all aircraft fly the same crooked beam, there is no problem. However, when an aircraft is on a false beam but believes it is on the correct beam, the results can be disastrous. Performing maneuvers while observing the received signal helped pilots identify these false beams. What was not needed was yet another complicated maneuver to fly the radio-navigation aid. What was sorely needed was a foolproof radio-navigation aid.

Various solutions to the problems of the AN ranges were tried out. One system used the same low frequencies but provided the ability to select any course bearing. This system was primarily intended for ship use, where the low-frequency waves are well behaved traveling over only seawater. Over land, this system suffered all of the problems of the AN range.

Some ultrahigh-frequency-range solutions were considered. What were called ultrahigh frequencies in the mid-1930s are now called very high frequencies, or VHF. The shorter wavelength of the higher frequencies would solve many of the problems of the AN range, such as the shore effect, the night effect, and false beams. A new antenna system, the Alford loop, invented in 1938, played an important role in these early experiments with shortwave navigation systems. This antenna system produced beams that were straighter and less affected by terrain than any other antenna system, but the experiments were nothing more than implementing the same AN range with shorter wavelengths. The system would provide only four courses and still required complex maneuvers to properly intercept a beam.

Improvements were also made to the 75-MHz markers by using improved antennas, particularly the Yagi-Uda antenna. Experiments were carried out on microwave marker beacons, culminating with a demonstration at Boston's Logan airport in 1941. Microwaves in 1941 meant 700 MHz, which is only an ultrahigh frequency (UHF) today. Indeed, the short wavelengths produced by the 700-MHz markers allowed for very narrow beams, but the technology to make affordable airborne receivers was not available at the time.

When the AN range became the backbone of the federal airways system, the system served many of the country's major airports. Instrument flight still had its problems. First, air-transport aircraft were not pressurized; thus all air-craft were restricted to altitudes below ten thousand feet. Because aircraft could not achieve the altitude necessary to fly over the weather, they often flew long trips completely in the clouds and in instrument conditions. This was downright dangerous in the days before radar. In addition, missing from the scene was a landing-navigation aid to guide aircraft to the runway with the aircraft at the runway heading. Ideally, the landing system would be in three dimensions. Remember, vertical guidance was nonexistent in Doolittle's epic 1929 flight, when the airplane was simply flown toward the ground. In the 1930s there was still no landing system. Although air transportation had vastly improved from the 1920s, in the late 1930s the chance of a flight's being can-celled due to weather was great.

Two important systems that are still in use today came out of engineer-ing programs started in the late 1930s. The first was an en-route system called the VHF omnirange, or VOR, system. The second system was the instrument landing system, or ILS. The Radio Corporation of America was developing an omnirange system using frequencies near 90 MHz (or a wavelength of three meters). Just before World War II there were only a few experimental receivers, but the system's performance was very encouraging.

The VOR, sometimes called "omni," was conceived to address the ma-jor problems associated with the AN range. First, the frequency range of the new navigation system is in the VHF part of the radio spectrum. Interference from atmospheric sources such as storms decreases with increasing frequency, and above 100 MHz the interference is minimal. There are no bent beams due to soil conductivity. There is no night effect as the signals travel nearly line of sight and are not propagated through the ionosphere. The VOR operates in the frequency range from 108 MHz to 118 MHz. Second, it operates with a vi-sual indicator rather than requiring the crew to listen to Morse characters. The indicator shows how far the aircraft is from intercepting the beam. This means that simple maneuvers are required to properly intercept the beam.

Third, the navigation aid provides any course orientation the flight crew desires. They can simply dial in the desired course and fly it. If the crew intercepts an unknown VOR course, they can determine that course and fly to that VOR station. The VOR visual indicator shows the deviation from the selected course as a visual aid to remain on course. The indicator shows when the aircraft has flown over the VOR station. Above the VOR station the pilot encounters a "cone of confusion" similar to the "cone of silence" found above the AN range. There is no cone of silence because the VOR does not require listening to the signal to navigate. However, the indicator flutters for a short time before registering the station passage, which serves as a position indication.

The omni has a Morse-code identifier to ensure the correct station is tuned on the receiver and that the signal is strong and usable. But other than an initial verification, there is no need to continually listen to the VOR receiver. What if the signal becomes weak or somehow corrupted? How would the aircrew know this?

A feature called a "flag," which would become the first of its type, addresses this situation. An electronic circuit monitors key signals in the VOR receiver. If one or more of these signals are not correct, a small flag appears in the indicator. This flag is typically bright orange or red and displays the word "off." It appears virtually in the center of the indicator and warns the flight crew that the instrument should not be used. This was the first use of "signal integrity monitoring," which is still a part of modern aviation electronics. Not only are the circuits and signals of the airborne equipment monitored, but the ground stations are as well. If a VOR ground transmitter has a problem, signal-integrity-monitoring equipment causes the station to be removed from the air. This causes the flags in all VOR receivers using that ground station to show "off," and the pilots find alternate navigation methods.

It was becoming clear that when aircraft were relying on radio navigation in instrument meteorological conditions (IMC), the safety of the flight could hinge totally on the ground and airborne electronics. Flying an aircraft with impaired electronic signals is on a par with flying the aircraft with an impaired crew. Strict regulations were enacted for the latter, and the same happened for the electronic systems. This led to a system of "technical standard orders," or TSOs, for airborne equipment, "national standards" for ground stations, and a host of controlling documents to ensure the safe operation of electronic equipment both airborne and on the ground.

To better understand the operation of the omni, imagine the rotating beacon at an airport. The alternate flashing of the white and green lights for a civilian airport is easily seen at quite a distance on a clear night. Let us assume we somehow knew when the green lens of the beacon was pointed directly north. Let us also assume the light rotates at a specific rate, which we also know. We could determine the position of the light by simply determining the elapsed time from when the light was pointed north and when we saw the green flash.

Let us now introduce a blue flashing light that is visible anywhere around the tower. This would be similar to the flashing obstruction lights on a radio tower to warn aircraft of the tower's presence. This light flashes when the rotating beacon is pointed north. As an example, say the light turns at a rate of one revolution every ten seconds. We see the blue light flash, and four seconds elapse before we see the green of the beacon. In that time the beacon has rotated four-tenths of 360 degrees, or 144 degrees.

The VOR uses the same type of signals but with radio waves rather than light. These signals can penetrate fog and clouds as if they were not there. The VOR station uses a rotating antenna pattern to produce the equivalent of the rotating beacon. The antenna does not physically rotate but is rotated electronically. Unlike the rotating light beacon, the radio signal rotates 30 times per second. When this signal is received in the aircraft, its strength increases and decreases 30 times per second due to the rotation. This permits a large number of observations, which ensure accuracy and an instantaneous response to changes in position. The VOR station also transmits a signal from an omnidirectional antenna, which contains a signal that also increases and decreases 30 times per second and simulates the blue light in our example. This signal is called the *reference*. The comparable signal from the rotating antenna is called the *variable*. The initial development of the VOR took place in the late 1930s and continued during World War II.

Improving Radio Communications

In addition to the VOR, VHF communications channels were assigned to the aviation services. Using VHF for airborne communications provided a number of significant advantages. First, the short wavelength, about 2.5 meters, permitted the use of full-sized, transmitting antennas on aircraft. This provided high efficiency for transmitting and permitted the use of lower-powered, airborne transmitters. A power of only a few watts would provide coverage to more than fifty nautical miles. The lack of ionospheric propagation meant much less interference than the longer wavelengths produced.

In the late 1930s most airborne communications were in the high-frequency (HF) range, generally from three to fifteen MHz. By this time the technology was reasonably mature and the communications, reliable. Radio amateurs had used this frequency range for a number of years, so the propagation characteristics were very well known. People were beginning to understand the ionosphere and the atmosphere in general. However, the high-frequency range had problems, too, but the technology to transmit and receive very short waves was just not ready yet.

One of the first radio communications sets was made by a small company in northern New Jersey—Aircraft Radio Corporation. This company was no stranger to aviation. The company was once a division of Radio Frequency Laboratories, who made the radio navigation equipment for Doolittle's blind flight. The aircraft radio division of Radio Frequency Laboratories became so successful that it grew to be larger than the parent company and was reorganized as Aircraft Radio Corporation and then absorbed its parent. By 1933 Aircraft Radio Corporation navigation and communications equipment had been installed in both U.S. Army and Navy fighter aircraft.

When war started in Europe, Aircraft Radio had a design for an entire set of communications equipment. It is important to understand that a communications set for an aircraft is not a simple, single piece of equipment. In the case of the Aircraft Radio communications set, a typical installation included a rack of two transmitters, a rack with three receivers, an antenna

The cockpit of a DC-3, vintage 1939, showing the first officer with a radio microphone, indicating the use of two-way radio.

changeover box, an antenna current meter, and a power supply. These racks were usually mounted just aft of the cockpit to keep the interconnecting wires to a minimum and to allow access to the equipment by the radio operators. These units were remotely controlled from the cockpit with a mix of mechanical and electrical connections. There were control boxes capable of tuning the receivers but not the transmitters. The transmitters were set either on the ground or during flight by the radio operators. If the transmitter strayed from the desired frequency, the ground station would follow the transmitter's frequency. The control boxes were capable of feeding several sets of headphones, which was necessary because both pilot and copilot must be able to use the communications equipment. The control box also allowed the pilot and the copilot to communicate with each other and other crew members. This *interphone* is a very important system aboard aircraft.

Both handheld microphones and microphones that operate with oxygen masks were used. Few military aircraft of the period were pressurized, so oxygen masks were used extensively. Many different microphones were used, including a "lip" microphone that fastened just below the lip with an elastic band and a "throat" microphone that picked up vibrations from the larynx.

The entire set of communications equipment—called the *command set*—had the military designator ARC-5. This was one of the early uses of the new, standard, military nomenclature. "ARC" stood for Airborne Radio Communications. The first letter indicated the type of installation. The second letter indicated the type of equipment, and the third indicated the purpose. Other examples included the ART-13, which was an airborne radio transmitter. In some examples the number was preceded by "AN," which originally meant army/navy. For example, for the AN/VRC-12, the number refers to a set of equipment used for vehicular radio communications (VRC). The receiver for this set is the R-40/VRC-12. The system of nomenclature is still used, but in 1971 it was renamed the "joint electronics type designation system," or JETDS.

Nearly a million pieces of ARC-5 equipment were built, including transmitters, receivers, and their accessories. During World War II, the Aircraft Radio Corporation was housed in wooden buildings along a grass airstrip in Boonton Township, New Jersey. This was the same grass strip Jimmy Doolittle had often landed on when Radio Frequency Laboratories was installing radio equipment for the blind-flight demonstration. Although ARC had production facilities, the small factory could hardly provide the number of command sets that the war effort required. Larger production facilities had to take over production using a "build to print" process.

This was a common situation during the war. As an example, Cessna Aircraft Company built ailerons and rudders for Boeing. A large aircraft aileron is similar in size and construction to a small aircraft's wing and suited for Cessna. Although a host of factories turned out command set parts, perhaps the largest builder of ARC-5 equipment was Western Electric. This company was the manufacturing arm of the Bell Telephone Company, which during that time was a nationwide, legal monopoly. Western Electric's facilities were huge and played an important role in the war. In a later section we will discuss the role of Bell Laboratories, another arm of the Bell Telephone Company, in the development of radar.

The desire was strong to develop a VHF communications system, particularly after a block of VHF radio frequencies was set aside for communications. High-frequency technology had reached a high level of sophistication, thanks to the efforts of the radio amateurs. Their use of high-frequency, short waves created a large market for equipment, so a number of very successful companies were started, mostly by other radio amateurs, to produce communications equipment. Some of these were the Hallicrafters Company of Chicago, founded by Bill Halligan; Collins Radio Company, started by Arthur Collins in Cedar Rapids, Iowa; the Hammarlund Manufacturing Company of New York City; and the National Radio Company and James Millen Company, both in the small town of Malden, Massachusetts. A number of smaller companies had also begun making amateur radio equipment in the late 1930s, and most of these were enlisted to make military radio equipment. In many cases, the amateur equipment was used with few or no modifications. This was particularly true for the Collins equipment, which was very high quality and expensive. Collins produced specific designs for the military, such as the ART-13 previously mentioned. In the case of airborne equipment, the amateur equipment was not rugged enough for airborne use, so the circuits and components were repackaged to withstand the airborne environment.

The Collins ART-13 represents an important enhancement to airborne radio equipment. While testing airborne radio equipment during World War I, Edwin Armstrong remarked that the equipment had only a few controls. It was necessary not only to set the operating frequency of a communications radio but also to adjust the tuning of several stages of the transmitter. Every frequency change involved a two-handed, "tune and dip" process to maximize the transmitter's output power. Therefore, a radio operator was one of the crew members of many aircraft. Collins Radio Company had invented an "autotune" circuit, where various frequencies were tuned on the ground, allowing the settings to be automatically made while in flight. This eliminated the need for a radio operator on board the aircraft. From this beginning, Collins Radio, later Rockwell Collins, dominated the airborne communications market for years.

The U.S. Army began searching for a suitable, airborne, VHF communications set. Aircraft Radio proposed a VHF transmitter and receiver that could be manually tuned with a knob, much like a home radio receiver. The accuracy of the tuning of this transmitter and receiver was a function of the skill of the operator. In addition to the manual accuracy required for tuning the transmitter, the frequency of the transmitter was a function of its mechanical parts. These dimensions, unfortunately, would change slightly with temperature fluctuations. Moreover, when the transmitter's frequency changed, it would interfere with an adjacent channel. Aircraft Radio were masters at making mechanically stable transmitters, as is evident in their high-frequency command sets. They proposed to the army a manually tuned radio transmitter and receiver in chassis almost identical to the high-frequency command sets. Therefore, only a new transmitter and receiver needed to be designed that would fit into the existing racks and accessories.

Competition for the Aircraft Radio design came from Western Electric, whose proposal had two major differences. First, the Western Electric design was for a *transceiver,* that is, a transmitter and receiver in one box. This had been done before, but it still was a relatively new concept. Second, a very precise quartz crystal controlled the transmitter. Using a quartz crystal for frequency control was the most accurate method available, but the disadvantage of this design is that one plug-in crystal is required for each desired transmit frequency. However, if the required number of channels is small, say fewer than 10, all 10 crystals could be installed in the transmitter and selected with

The LeGrande Oregon interstate airway communications station, ca. 1940.

a switch. The problem was that there were more than a hundred available channels for military air-to-ground communications. A crystal-controlled, military vehicular radio transmitter of the time could hold 10 crystals, but the radio
came with a box of about 100 crystals, of which 10 were installed in the radio.
The remaining 90 crystals were kept at the depot, hangar, or motor pool and
installed when necessary. The Western Electric design was selected, and military aircraft were equipped with both high-frequency and VHF radio, a trend
that was carried over to civilian aircraft after the war.

Toward a Viable Airline Industry

With the installation of the AN range, the establishment of federal airways,
and a unified, airline radio network, the possibility of a viable airline industry
began to grow. The Boeing B-247 was introduced in 1933, and the Douglas
DC-2, the following year. The DC-2 could take fourteen passengers, as compared to the Boeing's ten, from New York to Chicago at 180 mph. The coast-
to-coast route was reduced from a two-day to a one-day trip by flying all night
using the available radio navigation. In 1934 the Douglas Sleeper Transport,
DST, an aircraft with fourteen sleeping berths, was introduced for the all-

night portion of the transcontinental trip. This aircraft was reconfigured in 1934 with twenty-one seats rather than berths, and the aircraft was designated the DC-3. This aircraft boasted a fifty-percent increase in passenger capability from the DC-2 but only a ten-percent increase in cost. This increase in capacity and the resulting reduction in operating costs was a major step in bringing about a profitable airline industry without government subsidies in the form of either airmail contracts or outright cash.

In the early 1930s the Douglas DC-2 and the Boeing B-247D were considerably faster than the old Ford tri-motors. When the tri-motors and the newer aircraft occupied the same airways, a dangerous mix of aircraft performance existed. Fast planes, slow planes, mail planes, passenger planes, and aircraft without radio transmitters—all of these were crowded into a few federal airways. For safe air travel, standards are required for procedures, protocols, performance of equipment, the sharing of radio frequencies, and so on. Developmental work was carried out, but there was no umbrella organization to ensure that all of the efforts aimed at a common goal.

Rex Martin, the assistant director of the Bureau of Air Commerce, wrote in 1935 that "Research in the technical development of aeronautical radio aids has been pursued independently by so many organizations in an intensive effort to anticipate and meet so many rapidly changing requirements that it is now apparent that some guiding agency is necessary for efficient coordination of effort." Martin invited interested parties—airlines, manufacturers, and military and government agencies—to a June 19, 1935, meeting at the Department of Commerce in Washington, D.C.

The result of this meeting was the creation of the Radio Technical Committee for Aeronautics, RTCA, a nonregulatory, guiding agency. The committee became a commission as the result of a 1941 reorganization. Also formed at the 1935 meeting were five subcommittees. This would be the organization of the RTCA until the present time. A subcommittee would be formed to tackle a particular problem and would include members with a specific interest in that particular problem. An additional category, the special committee, was created in 1946, when the RTCA took on a more international look. To date, nearly two hundred subcommittees and special committees have been formed.

It is historically interesting to note the subcommittees that were formed at the very beginning of the RTCA, because they tackled the most pressing radio navigation and communications problems of the period. The subcommittees dealt with antennas and the elimination of precipitation static; direction finding for aircraft; the use of AC power in aircraft; instrument landing devices; and aeronautical frequencies.

Precipitation static occurred when aircraft operated in or near precipitation. The low frequencies used by the AN ranges were particularly susceptible to this type of static, as previously discussed, and improved antenna designs were the likely candidate to reduce it. As it turned out, improved antenna designs did lessen interference from precipitation static, but discharging wicks, currently used on aircraft, were even more effective in eliminating it.

An important problem the RTCA subcommittees dealt with is standardization. In the case of radio direction finding, one manufacturer may display data in one format, while another manufacturer chooses a different format. The differences may be subtle, but the results are confusing and may be catastrophic. Of course, the perennial problem of a landing system was on the RTCA's agenda, too.

One of the first subcommittees was concerned with aeronautical frequencies. Radio was a rapidly growing field, and demands for portions of the radio spectrum were numerous. Present at the organizational meeting of the RTCA was E. K. Jett of the newly formed Federal Communications Commission. Jett was named chairman of this subcommittee. The Communications Act of 1934 established the current Federal Communications Commission (FCC) to replace the Federal Radio Commission. Although the RTCA was not a regulatory organization, representatives of government regulatory bodies were members of the RTCA.

The subcommittee on aeronautical frequencies recommended using frequencies in the 3-to-30-MHz band, but it realized that higher frequencies would be of great value to aviation. Therefore, another subcommittee was established for the purpose of examining aviation applications of ultrahigh frequencies. The FCC, in one of its first official actions, assigned a very generous band of VHF frequencies for airborne communications. Now that a suitable band of frequencies was set aside for airborne communications, it was time to decide how to use it.

It is clear from the subcommittees formed what was foremost on the minds of the airline operators and manufacturers in 1935: instrument landing and VHF. As it turned out, when a workable instrument landing system was perfected, it used VHF and even modern UHF frequencies.

The initiation of air-traffic control to the en-route phases of flight was in 1935. The airlines in cooperation with the Bureau of Air Commerce formulated procedures for en-route and terminal-area traffic control. By 1936 Newark, Cleveland, and Chicago airports were controlling the en-route phase of flight for aircraft intending to land at those airports. In July, 1936, the bureau took over the air-traffic control facilities and responsibilities. This marked the beginning of federal air-traffic control.

In 1937 the bureau embarked on a modernization program for the airways. The first significant change was the addition of voice capability to the existing AN ranges. Later ranges would all have voice capability for enhancement of air-traffic control. By the time the modernization program was completed in 1938, the federal airway system consisted of 231 range stations, 121 marker beacons, and twenty-five thousand miles of airways.

Throughout the 1930s, after the formation of Aeronautical Radio, more airlines were using two-way radio, and air traffic was growing. The high-frequency radio frequencies, 3 to 30 MHz, used by the airlines were becoming seriously crowded. Aeronautical Radio pushed for the use of frequencies higher than those currently in use in the VHF part of the radio spectrum. The problem that Aeronautical Radio faced was that no VHF communications equipment was available. Aeronautical Radio drew up the specifications for a suitable VHF communications system and, in conjunction with Bell Telephone Laboratories, designed the necessary transmitters and receivers.

In the summer of 1938, Aeronautical Radio conducted VHF communications trials on Transcontinental and Western Airlines (TWA) on its New-York-to-Pittsburgh run. The tests were successful, and Aeronautical Radio reported to the FCC that VHF communications would be ideal for airborne use.

The Beginning of World War II

On September 1, 1939, Germany invaded Poland, marking the beginning of the Second World War. The following year, when Germany occupied Norway, Denmark, Holland, Luxembourg, Belgium, and France, the immediate future of aviation was focused on a military effort. Many of the projects that involved the RTCA had military applications, such as the instrument landing system. At the eighth meeting of the RTCA, held twelve days after the invasion of Poland, an instrument landing system was demonstrated. This demonstration was two weeks short of being exactly a decade after Doolittle and Kelsey first demonstrated blind flight. Radio technology had improved immensely during that decade, and the most important difference between the guidance system in 1929 and that of 1939 was the wavelength that was used. The earlier system used wavelengths of about 650 meters, while the newer system used wavelengths of about 2 meters. The shorter wavelength was the key to landing-guidance accuracy. Because of the war, this system would languish for nearly another decade before widespread installation took place.

Realizing that successful radio navigation depended on frequency allocations with wavelengths on the order of two meters, the relevant subcommittee recommended the assignment of the 119–126-MHz band for use in aeronautical radio navigation. The advantages of using VHF frequencies for communications were understood as a result of the experimental VHF communications system used by TWA on its New-York-to-Pittsburgh route. Another RTCA subcommittee recommended the use of the 129–132-MHz band for communications. However, all of these recommendations and demonstrations of instrument landings were put on hold until after the war, and RTCA essentially stood down during the war.

Aeronautical Radio arranged for Western Electric Company to manufacture VHF communications equipment designated ARC-1, for ARINC radio communication number 1. It is only a coincidence that the nomenclature "ARC" would become the military standard for "airborne radio communications" equipment. The first production units were delivered for commercial airline use in 1941, just a few months before the Japanese attack on Pearl Harbor. The U.S. Navy, realizing the advantage of VHF communications, diverted the entire production line of ARC-1 radios for military use. Interestingly, ARINC was able to repurchase many ARC-1 units on the surplus market after the war and redirect the radio back to airline use once again.

Further Reading

Dachis, Chuck. *Radios by Hallicrafters.* Atglen, Penn.: Schiffer Publishing, Ltd., 1999.

Redfield, Holland L. *Instrument Flying and Radio Navigation.* New York: Ronald Press Company, 1944.

Reilly, H. V. Pat. *From the Balloon to the Moon: A Chronology of New Jersey's Amazing Aviation History.* Oradell, N.J.: H. V. Publishers, 1997.

Rodengen, Jeffrey L. *Legend of Cessna.* Fort Lauderdale: Write Stuff Enterprises, 1998.

Sandretto, Peter. *Principles of Aeronautical Radio Engineering.* New York: McGraw Hill, 1942.

U.S. Department of Transportation. *Federal Radio Navigation Plan.* Washington, D.C.: GPO. Controlled by the U.S. Coast Guard and updated every two years.

Chapter 4

The War Years

The potential of aircraft was not fully explored during World War I. England endured 103 air raids, with 4,800 casualties, of which 1,400 were fatal. Relative to the total number of war casualties, this figure is almost insignificant. However, the air raids occurred with no defense, and air power in future wars would not be so insignificant. It was clear to armies and navies throughout the world that airborne delivery of munitions must be perfected and methods of defense against airborne attacks, developed.

An effective navigation system is the key to accurate bombing. The bombing runs of the First World War were ineffective because of the lack of a navigation system. A radio-based, military navigation system has to be immune to jamming and should not require transmissions from the aircraft. Making transmissions from a warplane is very undesirable because it gives away the aircraft's position. Thus, the navigation system should provide reasonable position determination using only airborne receivers. Germany had developed several suitable, radio-navigation systems after the First World War and used them for bombing raids.

The navigation system developed for airways navigation, the AN range, had no value in the European theater because it required a network of ground stations with large antennas. Only four courses were available, and thus guidance is available only on those courses. Furthermore, the AN range did not provide position information but only kept aircraft on a specific course. If a bombing target did not lie on one of the AN courses, there would be no useful guidance. For bombing, the position of the aircraft relative to the target must be known in order to successfully deliver ordnance. Wartime navigation must either utilize a long-range system or be transportable and capable of being set up quickly and moved when necessary.

In the case of ground defense, the location of the aircraft in three dimensions is required to aim antiaircraft guns. Radio waves have always been the choice for this application because of their ability to easily penetrate clouds. Lower-frequency radio waves follow Earth's contour and thus allow detection of aircraft over the optical horizon.

On the AN range, a plane's position can be determined at the point where two courses intersect. If the courses could be adjusted so that two of them intersect at the target, the AN range could be used for bombing guidance. Unfortunately, AN range signals can be received on the ground, and any adjustment of the intersecting beam would tip off the enemy as to the target of the bombing run. A similar system used only one beam that contained an audio-frequency tone. The aircraft received the tone and then retransmitted it to the ground station. The time delay was measured, and thus the position of the aircraft on the beam was ascertained. When the aircraft reached the bomb-release point, it was given instructions from the ground. As discussed earlier, the significant disadvantage of this system is that the aircraft is required to transmit and thus reveals its position, destroying its cover. Reliable, long-distance navigation was clearly a critical requirement for the successful outcome of the war in Europe. More important was the requirement for an early-warning system for invading aircraft. Nowhere was this more important than in Great Britain.

The Need for Early Warning

Some of the earliest experiments with radio waves discovered that nearby objects affect the reception of radio waves. Early shipboard radio reception was affected by nearby ships. It was not difficult to surmise that radio waves were reflected from the ships and then interact with those waves that were not reflected. This well-known phenomenon of electromagnetic waves is called *interference.* Using reflected radio waves to detect objects through clouds and over the horizon was a concept almost as old as radio. Of course, the devil is in the details, and the very long wavelengths used in the early radio experiments were suited only for the crude detection of large ships. It would take decades to improve the state of the art to generate sufficiently short waves to detect aircraft with precision.

In the mid-1930s, well before Britain's formal entry into the war, fear of Germany prompted the British parliament to develop radio detection facilities to ring the entire British Isle. In 1935 British scientists developed a system that today is called radar, or radio detection and ranging. The British engineers called their system radio distance finding, or RDF. This is not to be confused with the term *radio direction finding,* also RDF, used extensively in the United States. The early British system did detect and range—or find the distance to the target—but compared to what radar would become in the future, that system was hardly radar.

The plan was to develop a chain of RDF sites extending from the Orkney Islands, at nearly the northernmost point of the nation, to the Isle of Wight, at the southernmost end of the English Channel. The nearly uninhabited Shetland Islands in the northernmost region had no eastern-front protection. The entire set of RDF stations, dubbed "chain home," consisted of twenty-one stations. For maximum range, each station had four towers of 250 feet each. The operating frequencies varied from 23 MHz to 30 MHz, which corresponds to wavelengths from 10 meters at the higher frequency to 13 meters at the lower.

The chain-home system's main purpose was to detect aircraft at a distance that would permit dispatching defender aircraft to engage the invading aircraft before they could drop their bombs. Detection systems such as chain home had a profound effect on the design of aircraft. Future aircraft had to be designed with methods of avoiding detection from RDF or radar.

Detection of incoming aircraft is an essential application of RDF, but the ability to install RDF in aircraft suggests other, very powerful applications. One such application is the airborne interceptor, or AI. This is primarily an air-to-air application. The use of such RDF ensures the effectiveness of fighter aircraft as well as all-weather and night operation. The other major application of airborne RDF is the detection of ships—air-to-surface vessel, or ASV. Actually, "surface vessel" is a bit misleading. The most important surface vessel that an ASV RDF can detect is a submarine (ideally, an almost-surfaced submarine) or, more precisely, a submarine periscope.

There are a number of significant challenges to installing RDF in aircraft. First is the wavelength. The size of an antenna is dictated by the transmitted or received wavelength. In other words, the antenna size is *directly proportional* to the wavelength. An antenna of a certain type that operates with the chain-home system on a wavelength of 10 meters would be one-tenth as long as the same type of antenna operated with a wavelength of 1 meter.

There were also challenges relative to the size and weight of the equipment. The airborne RDF should weigh no more than two hundred pounds with a volume of less than eight cubic feet. A typical chain-home transmitter filled an entire room and weighed several tons. An alternative would be to install only a receiver in an aircraft and use the ground-based transmitters, but even the chain-home receiver was too large for an aircraft. Of course, there was no hope of the chain-home transmitter ever being installed in an aircraft.

Nevertheless, an attempt was made to severely scale down an existing, ground RDF and to raise the frequency to 300 MHz, which is a 1-meter wavelength. No significant power had ever been generated at such a high frequency before. Eventually a 6.8-meter wavelength was accepted.

A number of very difficult challenges in the development of radar are clear if we understand how radar works. Imagine we are in a darkened room, and in this room there is a toy model of an aircraft hanging from a thin thread. Imagine also that this is a night fighter. We are given the task of finding the aircraft and directing defensive fire to it. The solution seems simple. Use a flashlight to illuminate the aircraft and aim the gun. But what if this room were the size of a large auditorium? Our puny flashlight would not provide sufficient illumination to see the model. There is a simple solution: Use one of the high-intensity spotlights used for stage productions at the auditorium.

The spotlight illuminates the performers on stage by providing a high-intensity, narrow "spot" of light. This separates one performer from the others, which is pretty much what radar should do. To find the aircraft hanging from the string, we must search with our spot beam until we find the aircraft. If the target is moving, we must track the aircraft to keep it illuminated.

The example of a toy aircraft hanging in a large auditorium represents an easier problem than the one the radar developers faced. The most-produced German bomber of the time, the Junkers JU-885, has a wingspan of twenty meters and an airspeed of 265 knots, which is 265 nautical miles per hour. If we wanted to provide a fifteen-minute warning, the detection would have to occur at a distance of 66 nautical miles, or 123 kilometers. Let us now scale that to our indoor example. The detection distance, 123 kilometers, is equal to 6,135 wingspans. If we were in an auditorium where the distance from the last row to the stage was 61 meters—and that would be a large auditorium—the wingspan of our model would have to be 1 centimeter, or about four-tenths of an inch. Imagine watching a show with actors less than half an inch high. The audience would not see a thing. The only way to watch such a show is to use binoculars. However, this scenario illustrates two of the problems of detecting aircraft with radar. First, it was necessary to use a powerful light source and to concentrate the light energy on the desired target. Second, it was necessary to amplify the light energy returned from the target using the lenses in the binoculars, which are essentially light concentrators. Thus, radar requires high transmitter power, concentration of the energy using a narrow-beam antenna, and high amplification of the returned signal from the target using the same antenna or another one.

These problems were only a few of those that the early pioneers of radar faced. One of the major hurdles was the use of shortwaves. Light is one of a number of forms of electromagnetic radiation. The shortest wavelengths of the electromagnetic spectrum are gamma rays. Longer wavelengths are X rays and then light. Heat waves are longer than light, and at the bottom, that is to say, the longest wavelength, is radio.

If we illuminated our one-centimeter aircraft model with light energy, the waves would be about one-half of a micrometer or one-half of a millionth of a meter. These waves are reflected in all different directions from the irregular surface of the one-centimeter model airplane. The energy that is reflected to the source is what we would see. What would happen if we were to illuminate our model airplane with one-meter radio waves? The waves are one hundred times longer than the model and would simply pass by the target.

The real target, of course, is not a miniature airplane but full-sized models. A one-meter wavelength would be suited for detecting reasonably large aircraft. It would not detect the periscope of a submerged sub or a rocket. It would not be able to resolve multiple targets. A short wavelength would discern a difference between a trio of fighters or a large bomber. In the simplest terms, the shorter the wavelength, the better the performance of the radar. Around World War II, a one-meter wavelength would have represented the maximum wavelength suitable for a serious radar, with ten centimeters being the optimum. Generating high-power, short wavelengths posed a lofty goal in the years preceding World War II.

Radar held promise as a very effective defensive and offensive tool, but there were many ways to render it ineffective. One that was employed very early in the development of radar was *jamming*, which is the deliberate transmission of an interfering signal to render the radar's receiver ineffective. If a radar signal is jammed from a targeted aircraft, the radar can home in on the jamming signal to direct weapons. This is one of the first examples of a *countermeasure*. Every weapon of war has a countermeasure. Radar is the electronic countermeasure, or ECM, for aircraft. Jamming is the electronic countermeasure for radar and the counter-countermeasure, ECCM, or perhaps EC^2M, for aircraft. Using the jamming signal to direct weapons would be an electronic counter-counter-countermeasure, or EC^3M. To what power may we raise the "C" in ECM? Could we somehow encode the jamming signal so that the radar would not be able to determine the source and would thus incorrectly locate the aircraft? The answer is yes, it can be so encoded, and that would be EC^4M. This game of electronic leapfrog has been going on since 1935, and the industry has lost track of the power to which "C" has been raised. To keep things simple, let us say that the electronic warfare industry consists of electronic warfare, EW, and ECM.

Escalation of War in Europe and the Magnetron

On September 3, 1939, Britain declared war on Germany. By August, 1940, the Nazi Luftwaffe had been bombing Liverpool for nearly six weeks. The Nazis carried out the air attacks to soften the British defenses before an all-out invasion. Because of its long 10–13 meter wavelength, the chain-home RDF, although capable of detecting incoming bombers, had insufficient resolution to direct antiaircraft fire. Aircraft detected by the chain-home system required a visual observation, and antiaircraft guns were fired visually.

Although not ideal, the chain-home detection, when followed up by visual antiaircraft fire, foiled many of the daytime raids. It was assumed that the Luftwaffe would switch to nighttime bombing runs to cut their losses. Thus, on September 7, 1940, the Luftwaffe, as expected, began nightly raids on London. With these nighttime raids, the chain-home RDF was no longer capable of providing complete protection from aircraft due to its long wavelength and the use of countermeasures. Because antiaircraft fire from ground batteries or from fighter aircraft was still visual, the nighttime sky gave the enemy bombers all the protection they needed. A shortwave radar that could automatically aim antiaircraft guns, day or night, was imperative to Britain's future. This device was called a *gun-laying radar*.

In the late 1930s the available vacuum tubes were not capable of generating high-power, short wavelengths. Engineers understood that vacuum tubes for higher frequencies require that the elements of the tube be as close as possible to each other and that the leads connecting the elements to the outside of the glass envelope be as short as possible. Strange-looking vacuum tubes appeared with the required connections poking out of every end of the tube. One successful vacuum tube for frequencies up to about 900 MHz was made by the Radio Corporation of America and was called an "acorn tube." This tube was about the size and shape of an acorn and thus the name. Obviously a tube this small was not going to generate thousands of watts, even for a short time. The closely spaced elements also arced over and destroyed the tube when the high voltages required for kilowatt-power levels were applied. A high-powered vacuum tube could be physically small if there were an effective method of removing the heat.

A vacuum tube that the General Electric Company invented eventually became the preferred source of high power for radar transmitters. The history of this tube is rather interesting, as its invention had nothing to do with the generation of short radio waves. This tube used magnetic fields to control the path of electrons, so the inventors called it a *magnetron*. Although the magnetron is still used for radar transmitters at the time of this writing, it is not the source of choice for new designs. However, more magnetrons are manufactured today than ever before—for use in microwave ovens.

The magnetron, which held so much promise for the success of shortwave radar, had evolved over a period of more than twenty-five years. The electron tube, or vacuum tube, operates with beams of negatively charged electrons. Lee De Forest invented the first electron tube capable of amplification in 1906. There were many applications for the De Forest device in telephony, instrumentation, and radio, to mention just a few. De Forest patented the tube—which was called an *audion*—and was poised to make a fortune licensing manufacturers to use the device. In the audion the electron beam was controlled by a metallic element of thin parallel wires called a *grid*. This was the basic design of the De Forest invention. As early as 1916 manufacturers of radio equipment were searching for alternate methods of controlling the electron beam in a vacuum tube to circumvent the De Forest patent.

Electron beams can also be controlled with magnetic fields. The General Electric Company used magnetic fields to control the electron beam. However, electrons in a magnetic field can be made to take circular paths, which leads to the possibility of generating radio-frequency energy of short wavelength. A magnetron designed specifically for generating shortwave energy, called the *split-anode magnetron*, was capable of producing about five watts of power. This power was respectable for the time but inadequate for successful radar.

In late 1939 the University of Birmingham, England, began research into the generation of shortwaves. The split-anode magnetron, when used to generate microwave energy, requires a resonant circuit. *Resonance* is the phenomenon that determines the frequency of oscillation of the device. In the case of the split-anode magnetron, the cavity was external to the vacuum tube, which produced a number of problems, the most important of which was lack of cooling. The Birmingham team opted to place the resonant cavity within the tube and to use multiple cavities. This resulted in more effective cooling and thus permitted higher power output and, because of the reduction of heating, a more stable output. The internal-resonant-cavity magnetron was investigated in Japan, Russia, and Europe. Although the magnetron was conceived in the United States, most of its developmental work was done in Britain. Desperate to improve its RDF performance and turn back the Nazi bombers, the British hoped the magnetron would be key to a radar-directed, antiaircraft gun.

To use radar to automatically direct antiaircraft guns, a wavelength of about ten centimeters is required. Wavelengths this short or shorter are called *microwaves,* using the prefix *micro,* which is from the Greek *mikro* or *micros,* meaning "small." Other than the cavity magnetron, no other device of the time even came close to the requirements for a 10-cm radar. The magnetron provided this short wavelength in sufficient power. The trick was to produce a 10-cm radar before Germany could produce a 10-cm jammer, the countermeasure. This is a perfect example of a technology war.

The Magnetron Comes to America

The British high command decided that the future of the country depended on the ability to mass-produce the magnetron and provide a network of true radar stations to thwart enemy aircraft. They decided to enlist the aid of the American electronics industry to produce the magnetron. The British factories did not have the capacity that American factories had; furthermore, British factories were favorite targets for Nazi bombers.

At first it was suggested that the secrets of the magnetron be traded for American secrets, particularly the Norden bombsight. The Americans were very reluctant to give up such hard-earned secrets. Finally Churchill reminded the Royal Air Force that the future of Britain was at stake and that if the price was the secret of the magnetron, then it was time to pay that price and get on with it.

In late August, 1940, a metal strongbox was carried from London to Washington. The box contained all of the drawings and test data for the magnetron and a working model. The intention of the U.S. government was to turn the magnetron project over to the National Defense Research Committee, NDRC. The NDRC was a civilian committee whose purpose was to develop the technology of the nation's industries and universities for the war effort in Europe or—should it become necessary—for the United States' own war effort. The NDRC was chaired by an imposing and powerful character, the likes of which may never be seen again: Vannevar Bush.

A group of British scientists and engineers accompanied the magnetron to Washington and set up a meeting with members of the NDRC. The meeting was rather cool at first. The Americans outlined the state of radar and microwave research in the United States. Unlike the "microwaves" used for navigation aids during the 1930s, these were real microwaves—ten centimeters and shorter. At one point the U.S. members admitted they pined for more power for their radar experiments. Immediately the British delegation carefully removed the cavity magnetron from its wooden box and informed the NDRC members that the device could produce more than ten thousand watts of 10-cm power! The cool meeting became very warm, as this was ten times more power than the best American microwave transmitter could muster. This meeting marked the beginning of one of the world's greatest research and development efforts.

One of the first steps was to power up the magnetron and demonstrate its claimed performance. The magnetron was taken to Whippany, New Jersey, to the Bell Telephone Laboratories, where to the surprise of those present it did not demonstrate a power output of ten kilowatts, but closer to fifteen kilowatts. Now there was no question; the magnetron would be produced in the United States to be used by the British and the other Allied forces.

The day after the Bell Labs' demonstration, October 7, 1940, there was a strange turn of events. The British representatives were summoned to the labs to explain themselves. The drawings they had brought to America clearly showed a magnetron with six cavities, or cylindrical holes, bored into the copper body of the device. When the prototype device was x-rayed, there were eight holes. After a quick trans-Atlantic telephone call to England it was revealed that of the first dozen prototypes, ten had 6 cavities, one had 7, and one had 8. The eight-cavity magnetron, which was not fully tested, was shipped to America by mistake. This also explained why the magnetron produced fifty percent more power than expected. The Bell Labs scientists decided that the eight-cavity magnetron demonstrated the performance necessary and should be "reverse engineered." Thus, new drawings were generated from the eight-cavity prototype so that it could be copied.

Rad Lab

The U.S. government decided to establish a laboratory for the purpose of developing radar equipment for the war and, most importantly, immediately incorporating the magnetron into a 10-cm radar. After some wrangling over whether the lab should be a part of Bell Labs or the Massachusetts Institute of Technology (MIT), it was decided to locate the radar labs at MIT. At first the new laboratory was christened the "microwave laboratory" but was later deliberately changed to a somewhat misleading "radiation laboratory." A period of aggressive recruitment followed to staff the "Rad Lab," and on November 12, 1940, Bell Labs delivered five magnetrons copied from the British working model. Although not selected as the site of the Rad Lab, Bell Labs was an important player in the effort.

Slowly a talented staff materialized, and, with the magnetron as catalyst, work proceeded to produce microwave radar equipment to halt the German air offensive. The goal of the Rad Lab was to have a working prototype of an airborne, microwave radar installed in a night fighter by spring, 1941.

The first aircraft to receive a Rad Lab, airborne interceptor radar was the P-61 night fighter. The unit was manufactured by the Western Electric Company and was designated the SCR-520. A later unit, a much lighter SCR-720, replaced this unit about a year later. When the airborne interceptor, AI, and ground-control radar began to inflict heavy losses on the Luftwaffe, which was attempting to bomb London, there was a reduction in the number of bombing runs. In late 1940, the British embarked on their own bombing runs to the mainland. Now under siege as England had been a few years earlier, the Luftwaffe recognized the need to provide their own, reliable, air-defense system. Although Germany did not possess the magnetron, they had other techniques for building mostly longer-wavelength systems.

One countermeasure that required no installation of equipment aboard an aircraft was known to the British as "window," or "chaff" in the United States. Chaff was strips of aluminum foil of a specific length so that they would resonate at the radar's wavelength. Resonance would mean the chaff was very responsive to a particular wavelength. When a target is illuminated by radar and that target is resonant to the illuminating energy, there is a very strong return signal. A handful of chaff can provide a radar return that is as strong as a large aircraft. When huge numbers of these strips of aluminum are released, their light weight causes the strips to take quite some time to settle to the ground. During that period of time, the radarscope is overloaded with what is called *clutter*—literally thousands of targets from the strips. This is like pulling a window shade down and preventing the radar from seeing beyond the window. This is the basis of the English term "window."

This countermeasure did not remain a secret for long. After the use of chaff, the ground is littered with millions of tiny strips of aluminum. Thus, it was not long before chaff was used on both sides. Technological advances in chaff were aimed toward very small chaff, virtually invisible to the naked eye, which means huge amounts of chaff could be carried aboard an aircraft without a serious weight penalty.

The counter-countermeasure to chaff is to change the wavelength of the radar so that the chaff is no longer resonant. This characteristic of radar is called "frequency agile" today. This only prompted the development of more sophisticated chaff that was resonant at several frequencies to provide more complete protection. This is an example of ECM/ECCM warfare at its best and is one of the reasons the Second World War was called the "Wizard War."

The "Doodlebug" Arrives

By June, 1944, things were beginning to look good for Britain and the war. Britain had integrated its early warning, chain-home radar with airborne-interceptor and ground-based, control radar into a moderately effective deterrent to airborne invasion. It had been nearly three years since any serious penetrations of the British Isle by the German Luftwaffe. The invasion of Normandy had taken place in early June, and there was hope the end of the war was near.

Then on June 13, 1944, a strange sound was heard: a loud buzzing like a giant bumblebee. The buzzing suddenly stopped, and after a few seconds a great explosion was heard. The world will never forget this day, as it marked the first visit of the V1 rocket. The device appeared as a twenty-five-foot "cigar" that could carry a two-thousand-pound warhead more than one hundred miles. It wasn't really a rocket, as it used an air-breathing, pulse-jet engine. The rocket flew at a relatively low altitude of only a few thousand feet and at an airspeed of nearly 300 knots. Its small size and low altitude made the V1 difficult to detect with radar and difficult to intercept with a fighter, as 300 knots was fast for a fighter at the time.

Fortunately, the "buzz bomb," or "doodlebug," as it was called, was not a precision bomber. Unlike airplane bombing raids that followed radio beams, the V1 was programmed to fly a specific distance and then aimed in the direction of the target, usually London, and sent on its way. A small propeller mounted on the nose of the rocket and connected to what amounted to an odometer determined distance. The V1 was not accurate, but it was pilotless and inexpensive to make. They could be launched by the hundreds or thousands. It did not matter if half of them missed their target. More importantly, when would there be a V2? How much more sophisticated would it be? A defense for the doodlebug was imperative. These concerns were prophetic because a V2 was indeed being developed.

The Rad Lab provided a prototype, microwave, early-warning radar set to detect the doodlebug. The chain-home, early-warning radar was not microwave and was best suited for early detection of bombers and fighters. The 10–13-meter wavelength was too long for early detection of the smaller, high-speed doodlebug. MIT Radiation Labs developed a microwave, early-warning radar to supplement the chain-home, early-warning system. The early-warning, microwave radar was installed at Hastings, near the Strait of Dover, and was operating by July 4, 1944. It quickly demonstrated its ability to detect buzz bombs at more than 130 miles, which is virtually the full range of the weapon. This implied that the radar could detect a doodlebug as it was launched. Because the key to success in the most modern, ballistic-missile defense system is launch-phase (sometimes called boost-phase) detection, this capability is crucial. Modern missiles have ranges much greater than that of the doodlebug, and launch-phase detection is typically made by orbiting satellites.

The early-warning radar was operated around the clock, and fighter planes "stood" guard at five or six thousand feet. When a buzz bomb was detected, the ground controllers waited until the V1 was about one-half mile behind the fighter and then ordered the interception. Because the doodlebug was faster than most fighters, the descent from six thousand feet was used to build airspeed to match the three hundred or so knots of the V1 to make the kill. Although the fighters under radar control could make kills with the buzz bomb, a barrage of buzz bombs could overload the fighter squadrons and allow some doodlebugs through.

Fortunately, the most sophisticated fire-control radar of the day, another product of the Rad Lab, was in England waiting to be shipped to the continent. Through an emissary, the desperate Churchill asked Washington for two hundred of the radar-controlled guns. General Eisenhower realized the seriousness of the situation and arranged for an immediate loan of twenty radar-controlled guns.

The SCR-584 radar-controlled guns, the pride of the Rad Lab, controlled a battery of four guns with unparalleled precision. The radar fed information into the Bell Labs–designed predictor, which determined where to aim the gun based on where the aircraft was then and where it would be when the artillery round met up with it. The predictor was an electromechanical monster that performed mathematical computations. Today this device is known as a *fire-control computer.* After the war the fire-control computer became an all-electronic, analog computer and finally a sophisticated digital computer. Thus, the English island was protected with early warning and fire-control radar against air attack from bombers and unmanned missiles.

The Proximity Fuze

The guns employed another innovation, the *proximity fuze,* which was the ultimate in gunnery at the time. Early fuzes for bombs and artillery rounds detonated the round either on contact after a time delay or at a particular altitude. The proximity fuze could be set to detonate when it approached a set distance from the target. This was particularly important when the target was fast moving, such as an aircraft, or fast moving and small, such as the doodlebug.

The proximity fuze was the first electronic fuze to be used on artillery rounds. It worked with radio signals and to a certain extent is a radar device.

This was not the first proximity fuze ever made. There were attempts—some successful—at using optical sensors, sound, and magnetic fields. A magnetic proximity fuze had been used for naval mines. The radio proximity fuze was the best, all-around technique for a proximity fuze. Although it had a rather simple circuit, the challenge was to design an electronic device that could withstand the enormous acceleration of being shot from a gun.

The acceleration experienced by the fuze exceeded fifteen thousand g's. Practically no conventional electronic component could survive launch from a gun, and the least likely was the vacuum tube. Special vacuum tubes were developed and produced in huge numbers. Near the end of the war, Sylvania, the major supplier of ruggedized vacuum tubes for fuzes was producing 400,000 tubes per day, with plans for increasing production to 525,000. By comparison, the total vacuum-tube production in the United States at that time, excluding those for the proximity fuze, by all manufacturers and all types was 600,000 per day.

The secrecy given to the proximity fuze, known by the British as the VT (variable time) fuze, was as great as or even greater than that afforded radar research. The U.S. Army's code name for the proximity fuze was "POZIT." The fuze went into production in late 1942 and was first used in battle in the Pacific theater in January, 1943. The fuze was first used on naval guns for a very important reason: Unexploded rounds would not be salvageable for inspection by the enemy. This was particularly true in the deep waters of the Pacific. This ban on land use lasted for more than a year.

The effect of the proximity fuze was spectacular according to a paper presented in October, 1945, by Grant Hector, who was involved in the development of the proximity fuze. Hector stated that no battleships or light cruisers were lost due to air attack from the inception of the fuze to the end of the war. Only one carrier and one heavy cruiser were lost.

Bombing with Precision

The earliest experiments with bombing about the time of World War I involved the dropping of bombs over the side of aircraft by hand. Not surprisingly, most of the bombs missed their target. To have any success at all, bombing had to be done from such a low altitude that aircraft could be shot down with an infantry soldier's rifle. There was a clear need for a device for adjusting the flight path of an aircraft and timing the release of the bombs. The first device—called a bombsight—to meet this need was designed in 1920 for the U.S. Navy by Carl Norden, whose name would be forever associated with the device. The device was a gyro-stabilized, mostly optical device called stabilized approach bombing equipment, or SABE. To accurately drop a bomb on a target, a number of parameters must be considered. First, the location of the aircraft relative to the target must be known. Second, the bomb's characteristics must be known. This includes the glide characteristics (that is, how far the bomb drops for every increment of forward distance). Bombs that are aerodynamic and include fins that orient it can have a much better glide ratio than those that are fat and have a lot of wind resistance. Third, the aircraft's airspeed and ground speed must be known. This is because the aircraft's position relative to the target is referenced to the ground, while the actual path of the bomb is through the air. Fourth, the aircraft's altitude and the target's altitude must be known. Finally, the wind's velocity and direction must be known.

The bombsight had a long period of development from the 1920s to the beginning of World War II. The early device was essentially a mechanical analog computer, but it was the genesis of the modern electronic weapons systems found in all fighters and bombers. The early device was incredibly complex and contained almost twenty-three hundred different parts. As more data were automatically gathered (rather than being entered by hand by the bombardier), the bombsight acquired electrical devices. To call the World War II bombsight "electronic" would be quite a stretch, but it was definitely "electric." In 1944 the Norden bombsight was interfaced to the AN/AP 27 radio set and gained the capability to home in on radio signals. Shortly after the war, the bombsight was interfaced to airborne radars, which was a big step in the development of modern airborne weapons systems. With the later add-ons, the Norden bombsight was used in decreasing numbers up to the mid-1960s—a remarkable forty-year run.

The Norden bombsight was one of the most important secrets of the war. As discussed previously, the bombsight's plans were the suggested price for the secrets of the magnetron. The offer was refused, and the Norden secret was deemed too high a price to pay for anything. The British lack of a competent bombsight also figured heavily in their bombing strategies. The British approach to bombing was high-altitude nighttime raids. These altitudes were well above antiaircraft fire, and, since the raids were conducted at night, the German fighter aircraft, which lacked radar, were at a disadvantage. It was virtually impossible to spot the intended targets, and the long-distance navigation of the time was inaccurate. To compensate, the British bomb raids dropped huge amounts of ordnance in what was called "carpet bombing." Very little of the bombs met any important targets, however, and more civilians and farm animals were killed than damage was done to military installations. However, bombs occasionally met their mark, and, since so much ordnance was dropped, the technique was mildly effective.

The Allied strategy was different. Allied bombing raids took place during daylight hours. Like the British, high altitudes were ordinarily preferred in order to reduce casualties from antiaircraft fire, but the bombers had the advantage of daylight and the Norden bombsight. The Allied strategy was more vulnerable to attack by fighters, however, and tended to have a high casualty rate, but the precision bombing was more effective than carpet bombing.

Friend or Foe?

On a radar screen, all aircraft look alike. A green radar blip may suggest the size of the target, but it does not show whether the aircraft is friendly or hostile. A system was developed in 1943 called Identification Friend or Foe, or IFF. This system was the first use of secondary radar. The system is "secondary" in that it operates in conjunction with primary radar.

The IFF system, although designated a radar system, is actually a communications system. Equipment called a *transponder* is installed in aircraft to be identified by the IFF. The transponder contains a receiver and a transmitter and automatically transmits when it receives a properly encoded signal. A signal called an *interrogation* is transmitted from the radar site to the aircraft within view of that radar site. The signals are encoded in one of several different "modes," and aircraft receiving these interrogations transmit a reply. The data contained in a reply are determined by the mode of the interrogation. In addition to identity, which is mode 3, an IFF transponder can provide other data from the aircraft, most notably altitude, or mode 1.

Having a transponder aboard an aircraft opens up the possibility of hostile forces' receiving the transponder signals and homing in on these transmissions. Both the Allied forces and the Luftwaffe used secondary radar, and it wasn't long before the transponders were used for homing. IFF is still used today in both civilian and military applications. However, the modern transponder is a more sophisticated device and much more resistant to countermeasures.

Before the end of the war the Rad Lab had produced an entire spectrum of radar systems that included early-warning radar, precision-approach radar, fire-control radar, airborne-interceptor radar, and IFF. The Rad Lab actually produced much more than radar designs and hardware. Practically every radar system the Rad Lab designed involved aircraft, which included flying bombs and missiles. A notable exception was short-range, surface radar for detecting ships.

When the Rad Lab was formed, there was no technology for the development of microwave equipment. Microwave measurements are very different from normal electrical measurements. Sometimes microwave energy does not even travel through wires but travels as guided waves through hollow rectangular "pipes" called *waveguides*. There was no test equipment available on the market that would interface with waveguides. Microwave measurements produced results that differed from the accepted norm in electrical measurements. Rather than volts and amperes, microwave measurements were given in terms of forward and reverse power, reflection coefficients, and standing-wave ratios. A new language was used to describe microwave behavior. An entirely new methodology was required for microwave measurements that included both the measuring equipment and method of displaying parameters.

Not only did the Rad Lab invent modern radar, it also invented all of the tools necessary to work with microwaves. This was an absolutely phenomenal achievement. A twenty-seven-volume set of reference books known as the *Radiation Laboratory Series* was published immediately after the cessation of hostilities and served as the major reference for microwave engineers for decades after the war. The material in the set was suited not only for radar applications but also for all types of microwave engineering, which included other aviation applications.

The size of the Rad Lab was reduced after the war, but it was not completely shut down. Some of the staff remained to publish the Rad Lab series of reference books while others continued to develop radar applications. Two radar systems developed by the Rad Lab after the war were air-traffic control radar and weather radar. The Rad Lab also modified old military radar sets for space research.

Long-Distance Navigation

The Rad Lab is well known for its pioneering work on radar, but, in addition to radar, the Rad Lab had another high-priority assignment: the development of a long-range navigation system.

As we have discussed, an accurate, long-range system would have permitted nighttime bombing raids and resulted in reduced aircrew casualties. Both the Allies and Germany experimented with several systems. One German system was called Knickebein, or "bent leg." This system, as most other experimental systems, relied on radio beams that grew wider as the distance to the transmitter increased. The German Knickebein got its name from the fact that the system required two intersecting beams that formed the shape of a bent leg when plotted on a map. The countermeasure for Knickebein, which the British codenamed "headache," was jamming, which was appropriately codenamed "aspirin."

Alfred Loomis of the Rad Lab proposed a system shortly after the lab was formed in October, 1940. The proposal was for a "hyperbolic" technique and was based on an earlier system, Gee, developed by the British. Dubbed LORAN, for long-range navigation, the system provides position information.

Of the navigation systems discussed thus far, the nondirectional beacon gave heading information, and the AN range and VOR provided course information. An aircraft could home to the beacon, but the navigation system did not allow the selection of a specific course. If there were a brisk crosswind, the track would not be a straight line.

Relative to determining position, sometimes called a *fix,* or simply "where am I?" the nondirectional beacon gave no information other than "you are close enough to the station to receive a signal." Some indication of the actual distance could be obtained from the signal strength, but this was far from accurate. When flying directly over a beacon station, the signal strength would suddenly decrease in what is called the "cone of silence." This was a relatively good position fix, and thus beacons were placed at strategic locations such as airports. This is true of other beacons such as the marker beacon used with the later instrument-landing system and the airway beacons used with the AN range. These low-powered transmitters could be received only at short distances. Unlike beacons, the AN range and VOR provided a line of position, or LOP. These LOPs were radio beams and thus straight lines passing through the radio navigation station.

If an aircrew had two lines of position that intersected at one point, the LOPs would provide a position fix. This was used for bombing guidance and was the technique employed in several bombing-guidance systems, including Knickebein. In the case of the AN range, because there were only four courses, the LOPs seldom intersected.

This is not the case with VOR. First, there is an infinite number of courses from a VOR ground station, and the stations are placed closer together. Thus any aircraft in the coverage area of two VORs can determine a fix. At the time the Rad Lab was formed, VOR was only experimental. Furthermore, the short (about three-meter) wavelength used for the VOR propagates via straight lines, which is necessary for the successful operation of the system. In this case, however, "straight line" means the signal will not follow the curvature of the Earth and therefore will not provide the needed, over-the-horizon coverage to provide guidance deep into enemy territory.

A long-range navigation system would provide guidance not only for aircraft but also ships at sea. Since radio stations cannot be located every few hundred miles in the oceans of the world, a long-range navigation system must use signals that propagate many hundreds of miles.

The LORAN system required a minimum of three synchronized transmitters that transmitted pulsed signals that propagated in expanding circles. These transmitters transmitted in sequence so that the first transmitter (called the "master") would always be received before the other stations (called "auxiliaries"). A master and its auxiliaries were called a LORAN "chain." The chains were independent and covered a large geographic area. By measuring the time difference between any two received signals, a hyperbolic line of position could be determined. With three stations, three hyperbolas could be determined: one from each time-delay measurement. The three time delays are from the master and the two auxiliaries and between the auxiliaries. The receiver is located where the three lines of position intersect, which is a single point. A special chart on which the LORAN lines of position are overlaid on the normal geographical features is used to plot position. The first LORAN chain was installed in early 1942, and tests were conduced in June of that year. The frequency range chosen was the 160-meter, amateur-radio band confiscated from the now-silenced radio amateurs.

To be successful, any navigation system must have some resistance to countermeasures. The major advantage of the LORAN system was the use of enormous transmitter output power and multiple frequencies. Of course the countermeasures are enormous jamming power and multiple frequencies.

Several somewhat successful attempts were made to use LORAN for aircraft during the war. The early LORAN sets required manual manipulation of controls while observing pulses on an oscilloscope and then plotting position on a LORAN chart. This was not suited for smaller aircraft, but long-range bombers and other large aircraft could and did use LORAN. By the end of the war, there were about twenty LORAN chains providing coverage to almost a third of the Earth's surface.

One of the major problems of LORAN was interference from sky waves. The preferred wave from LORAN is the ground wave, which follows the curvature of the Earth. Sky waves, which are returned from the ionosphere, do not follow a predictable path and also interfere with the ground wave. This was the same phenomenon that caused the night effect in the earlier AN range. Another problem with LORAN is that there is considerable atmospheric noise in the frequency region of 1.7 MHz, and LORAN stations use very high power to overcome the noise. Atmospheric noise is also a major problem that afflicted the AN range and low-frequency homing beacons. Yet, the high transmitter power was the countermeasure to jamming.

The modern successor to LORAN, called LORAN-C, requires no manual manipulation as a microprocessor makes all of the necessary calculations. Although LORAN saw little use in aircraft, it is still an important phase in the development of aviation. The later LORAN-C version was adapted for airborne use in the 1970s, some thirty years after the inception of the original LORAN. The introduction of LORAN-C in aircraft marked a huge increase in aircraft navigation capability, particularly for small aircraft.

Before the war there were only a few radio-based navigation systems in the world. After the war there were approximately ten navigation systems that had either been fully installed and operational or experimental systems that were completed but not commissioned. The fully operational systems included LORAN, Decca, Consol, Gee, Oboe, Shoran, Rebecca/Eureka, and VOR. An English invention, the Gee system was a hyperbolic navigation system. Gee was similar to LORAN in that Gee charts were used, but the significant difference was that Gee used wavelengths that were ten to fifty times shorter than those LORAN used. Potentially, Gee could produce a more precise, reliable fix than LORAN but not the long range. The short wavelength used for Gee permitted lower transmitter power and freedom from ionospherically propagated interference but limited the range to line of sight. The range could be better than one hundred nautical miles for a high-flying bomber but was considerably less at lower altitudes. For surface vehicles and ships Gee had very limited range.

Of the ten systems, only LORAN, VOR, and Decca survived for any significant time after the war. Only VOR survives at the time of this writing, while LORAN survives in its modern form, LORAN-C.

The Ongoing Need for a Landing System

The end of the war marked more than fifteen years since Doolittle demonstrated blind flight, and there still was no landing system in widespread use. Some work had occurred in the late 1930s on what eventually became the standard instrument landing system, but that system was not suited for temporary installation at a landing zone or on an aircraft carrier.

One attempt to guide aircraft during instrument conditions used a ground-based radar called ground-controlled approach, or GCA. The system never became popular for a number of reasons. First, a ground-based radar and controller are required for the approach. Second, feedback to control the aircraft is by voice communications, which are very slow and prone to errors. When multiple aircraft are on approach, the system becomes so prone to errors that it is unworkable. The system was used only in emergencies or when an airfield or carrier was unexpectedly socked in by weather. The GCA radar scanned both azimuth and elevation and provided a three-dimensional picture for the ground or ship controller. These systems were relatively short range and used short wavelengths.

In spite of the need for a landing system after the war, radar-assisted approaches were never used for civilian aircraft except in emergencies. Shortly after the war, the instrument landing system, ILS, was perfected and installed in large numbers. The ILS provided accurate visual guidance for landing without turning control of the aircraft over to the ground controller. A precision-approach radar, PAR, is still used by the military, however.

Signals Intelligence

During the war numerous radio-based systems for navigation, control, communications, and countermeasures were developed and put on the air. It was extremely important that military operatives be aware of what signals were on the air at any time. For signals longer than about 75 meters, propagation is by ground wave during the day and both ground wave and sky wave at night. These signals can be received by ground-based receivers at significant distances. For wavelengths between 75 meters and about 10 meters, propagation is variable, but there is frequent, long-distance propagation. For waves shorter than 10 meters, however, propagation is line of sight. The best platform for monitoring signals with wavelengths shorter than 10 meters, which included many important signals, was an aircraft. Ground "listening posts" were im-

Before the development of the instrument landing system following World War II, ground-controlled approach radars were used to enable aircraft to land under instrument conditions. This photograph shows a U.S. Army Air Corps AN/MPN-1 radar in 1943.

portant in the early days of World War II, but airborne signal reception was becoming increasingly necessary.

Military organizations recognized the importance of continual monitoring throughout the entire radio spectrum and analysis of the signals received. Everyone knows the stories of breaking enemy codes and intercepting secret messages. What is not as well known is the importance of receiving navigation information and jamming signals in electronic warfare. The modern term for these activities is "SIGINT," which stands for "signals intelligence."

Receivers that cover the longer-wavelength portion of the radio spectrum have been available ever since radio amateurs started using shortwaves. Because the military also used shortwaves, militarized receivers were in reasonable supply. For the very short wavelengths, very few receivers were available in the early days of World War II. One significant exception to this was the Hallicrafters S-27 UHF receiver. Its frequency range was 27–145 MHz, or wavelengths of 10 to 2 meters, which is VHF in modern terms. This receiver featured simple operation, AM and FM capability, and a signal-strength meter. There were competing receivers at the time, most notably the National Radio Company's 1-10 receiver. The model designation "1-10" implied the receiver operated between the wavelengths of one and ten meters, which is exactly the entire, modern VHF spectrum. Although the National receiver covered a broader frequency range than the Hallicrafters, the 1-10 had no FM capability, required a calibration chart for the frequency dial, had no S-meter, and was very difficult to use. Both the 1-10 and the S-27 used the RCA-manufactured acorn tubes described earlier in this chapter.

The S-27 was put into military use, but the receiver was not intended for airborne use. There was a pressing need for an airborne surveillance receiver that would be simple to use and cover a broader frequency range. As the frequencies used for radar and other electronic warfare were increasing, the 145-MHz upper limit of the S-27 was inadequate. To fill the need for a reconnaissance receiver suitable for airborne use, the AN/APR-4 was manufactured by a number of companies, including the venerable Aircraft Radio Corporation. The AN/APR-4 covered a very wide band of frequencies up to 1,000 MHz by using plug-in tuning units. This technique offered a number of advantages. First, different design schemes are used for different frequency ranges. That is, the components and designs for, say, 30 MHz are considerably different from those used for 1,000 MHz. Second, as technology advanced, the need for higher-frequency coverage could be met by designing a new tuning unit rather than an entire receiver because only the "front end" of a receiver must handle the higher frequencies.

However, the plug-in design had disadvantages, too. In order to change frequency bands, the operator was required to remove the tuning unit from the receiver, which was quite large, stow it away in a rack, and install a new unit. Still, one receiver and four tuning units made a lighter load than four receivers.

The AN/APR-4 had a serious flaw that became evident when the receiver was placed into service. The superheterodyne receiver requires an oscillator circuit, called the *local oscillator,* to convert the input to a lower frequency. An oscillator is a generator of radio-frequency energy. Remember, it was the invention of the oscillator that permitted pure "continuous waves" to replace the early spark transmitters. If the surveillance receiver's antenna radiates even a small amount of energy from the local oscillator, the receiver becomes a transmitter.

A small amount of energy from the AN/APR-4's oscillator could get coupled to the receiver's antenna and radiate. Enemy monitoring stations could receive this signal and determine not only the presence of a hostile aircraft but also the exact frequencies that aircraft was monitoring. This would permit the enemy to shut down operating radars or other radio systems or to transmit misleading signals to confuse the reconnaissance effort. When this weak link in the AN/APR-4 was discovered, modified tuning units with greatly reduced local oscillator radiation were provided.

The radiation of signals from an aircraft, including both those that are deliberately radiated and those that simply leak from operating equipment, is called the aircraft's *signature*. Unintentionally radiated signals, such as those from the local oscillator, represent a subset of SIGINT called URINT, for unintentionally radiated intelligence. Modern SIGNIT relies on very complex radiated signatures, so modern aviation electronics must be as free as possible of radiated signals. Some of the best-kept secrets of a modern military are the radiated signatures of airborne electronics systems. The signatures are so secret that in some cases the manufacturers must submit equipment they have manufactured to the military for testing. Not even the manufacturers are permitted to know how the equipment is tested or the signature of the equipment. The manufacturer only knows whether the equipment passes muster.

Electronic Components for Aircraft and Spacecraft

The war years brought about the visible electronic systems such as radar, the proximity fuze, and improved navigation, but there was also an unforeseen development. Electronic components improved immensely during the war years. Vacuum tubes became smaller and more rugged. Normally unimportant components were improved. The common capacitor, then called a *condenser*, would not survive the acceleration of the proximity fuze. Very rugged and inexpensive capacitors were developed. The cost of the capacitor was important because the fuze was an expendable item. These improved capacitors were then used in all systems.

As the military came to rely more on their electronic systems for successful campaigns, the potential failure of these systems became a critical issue. Engineering departments began to design for reliability and to investigate the physics of failure. They designed systems to provide not only a certain level of performance but also more reliable performance. To predict reliability, they used methods that analyzed the failure history of similar components. In the early days of radio, before the World War I, the major thrust of the electronics industry was to provide a fancy product to the public. Reliability was not especially important. If a radio failed and the repair involved more than plugging in a new tube, it was no catastrophe to send the radio to the dump. After all, it was an old model, and much had changed.

Another factor was the increasing complexity of electronic systems. For years, the standard, inexpensive radio had five tubes. A rather fancy one might have nine, while the top of the line would have perhaps twelve tubes. The more sophisticated, airborne electronics systems such as radars might have dozens of tubes. Vacuum tubes and other components made to operate in a simple five-tube radio would fail at an excessive rate in a fifty-tube radar set.

One of the first serious efforts to improve electronic reliability involved the design of repeaters for undersea telephone cable. These amplifiers were expected to operate twenty-four hours a day without failing. Since pulling up a submerged cable for repairs was a very expensive operation, it was important to design in reliability. The modern analogy is a satellite. Unlike the undersea telephone cable, satellites cannot be retrieved for repairs at any price. The most reliable electronic circuits ever designed are found in spacecraft.

Modern aircraft have benefited from more than fifty years of reliability engineering. Because of the enormous amount of critical electronics aboard modern aircraft, both military and civilian, reliability is more important than ever.

During the war, production efficiency of American factories, already the finest in the world, was improved, and aircraft rolled out by the thousands. Millions of electronics systems were produced. Aircraft by the end of the war had both VHF and HF communications, airborne radar, radio navigation, IFF, electronic countermeasures, and precision bombsights. Electronics would forevermore be an integral part of aircraft.

Further Reading

Baldwin, Ralph. *The Deadly Fuze: Secret Weapon of World War II.* San Rafael, Calif.: Presidio Press, 1980.

Brown, Louis. *A Radar History of World War II: Technical and Military Imperatives.* Philadelphia: Institute of Physics Publishing, 1999.

Buderi, Robert. *The Invention That Changed the World: How a Small Group of Radar Pioneers Won the Second World War and Launched a Technological Revolution.* New York: Simon and Schuster, 1996.

Bush, Vannevar. *Modern Arms and Free Men: A Discussion of the Role of Science in Preserving Democracy.* New York: Simon and Schuster, 1949.

Daso, Dik A. *Hap Arnold and the Evolution of American Air Power.* Washington, D.C.: Smithsonian Institution Press, 2000.

Dean, Francis H. *America's Army and Air Force Airplanes: 1918 to the Present.* Atglen, Penn.: Schiffer Publishing, 1997.

Hector, Grant. "The Radio Proximity Fuze." *Proceedings of the Radio Club of America,* March, 1946.

McFarland, Stephen D. *America's Pursuit of Precision Bombing, 1910–1945.* Washington, D.C.: Smithsonian Institution Press, 1995.

Pilots' Information File, 1944: The Authentic WWII Guidebook for Pilots and Flight Engineers. Atglen, Penn.: Schiffer Publishing, 1995.

Schoenfeld, Maxwell Philip. *Stalking the U-Boat: USAAF Offensive Antisubmarine Operations in World War II.* Washington, D.C.: Smithsonian Institution Press, 1995.

Stewart, Irvin. *Organizing Scientific Research for War: The Administrative History of the Office of Scientific Research and Development.* Boston: Little Brown, 1948.

Streetly, Martin. *Airborne Electronic Warfare: History, Techniques and Tactics.* London: Janes, 1988.

Chapter 5

The Return of Air Travel

A fter World War II, the growth of air travel was aided by the implementation of the technology advances made during the war. This included many applications of radar, pressurized aircraft, and improved navigation.

After the war, VOR appeared to be the most promising means of providing medium-range, en-route navigation. Before the war the FCC had assigned a range of VHF frequencies for airborne use, but with the increased demand for spectrum for new navigation systems, the agency was pressured to assign more VHF spectrum for airborne use. The radio amateurs had the frequencies from 112 to 116 MHz, which was very close to the existing airborne VHF assignment. After World War II, the scenario of World War I was repeated. The FCC was not allowing the amateurs back on the air. When they were finally allowed back, the amateurs found that two of their radio bands had been eliminated without compensation—one for the new VOR and another for television broadcasting. It took an act of Congress to force the FCC to compensate the radio amateurs. The amateurs received a 4-MHz band of frequencies above the new aviation VHF assignment as well as two-thirds of the new television channel 1. To this day there is no channel 1 on U.S. VHF television receivers.

Enormous advances in very shortwave technology were made during the war. This allowed the VOR to become a highly advanced navigation aid. After more than a half century of operation, the VOR is still the backbone of the airways system. The Federal Radio Navigation Plan calls for decommissioning the system by 2015, with most navigation being performed by satellite-based systems by that date. This will result in a seventy-year reign of VOR as a major navigation aid. However, there is some speculation that the VOR will remain as a backup system and that decommissioning may not take place until decades later, if ever.

Creating International Standards

As the United States and their allies worked together during the war to defeat their enemies, it was becoming evident that standardization would be required for an international aviation business. Before the war each country developed its own unique radio navigation system. In the United States the AN range was the en-route radio navigation system. The English had Decca, and the Germans had several systems. In addition to radio navigation aids, there were other needs for standardization. For example, how should an airport appear? How are the runways marked? Where is the airport's light beacon placed? What are the procedures for approach and landing? What language should be used for voice communications?

In November, 1944, the United States and its allies met in Chicago to form the International Civil Aviation Organization, or ICAO. Like all good aviation acronyms, ICAO has a pronunciation, which is "eye-kay-oh." ICAO has since become a part of the United Nations and is headquartered in Montreal, Canada.

After World War II, air-traffic control was again due for a major overhaul. The volume of air traffic that occurred before the war would not only return and but also increase at a rate never before imagined. The new air-traffic control system would first be a radio-based system and then, in a few years, radar based. Before the war the air-traffic control system was radio and telephone based or teleprinter based, using airways defined by AN range stations. The systems invented during the war years—radar, VHF communications, ILS, and VOR navigation—would set the stage for the modern air-traffic control (ATC) system. The new air-traffic control system would have to handle the new jet aircraft that flew at much higher altitudes and airspeeds than aircraft up to that time.

The postwar period was one of considerable modernization of the federal airways. Although the VOR had been developed before the war, very few installations were made in the United States because of the war. Installing VOR stations and using them as anchors for the new federal airways—now called "victor" airways—was the first step toward the modern airways system. The name "victor" comes from the ICAO's standard phonetic alphabet. Since so many English letters—c, b, d, e, g, v, and so on—sound similar and can easily be confused, words are used when reciting letters. The preceding sequence would be "charlie, bravo, delta, echo, golf," and "victor."

Because of the VOR's ability to provide virtually any course line, VOR stations could be installed wherever they would make sense, such as at major airports. Then the VORs could be connected to other VORs to make a victor airway. The VOR was not limited to only four courses as the AN range was. VOR courses had no restrictions on the minimum angular separation. Thus the postwar airways map included a much larger number of airways with more convenient placement.

The Elusive Landing System Finally Arrives

In spite of all of the advances in radio technology during the war, hostilities were over before a good, radio-based landing system could be installed for military or civilian use. The system that ultimately emerged included three radio elements: the localizer for horizontal guidance, the glide slope for vertical guidance, and beacons for position determination. The ILS operated on well-formed radio beams that were shaped by precision antennas. One beam provided lateral guidance, while a second provided vertical guidance, and fan-shaped beams provided a marker function. The localizer operated in the VHF portion of the radio spectrum, while the glide slope operated in the UHF portion. The UHF range goes from 300 MHz to 3,000 MHz. The wavelength of the 330-MHz glide-slope signal is ninety-one centimeters. This wavelength was seriously pushing the state of the radio art, but work had been progressing on the ILS when the war efforts took precedence. It took almost twenty years—from 1929, when Doolittle first demonstrated blind-flight landing, until after the war—before a civilian landing system became available for use.

A bit of explanation will help explain why it took so long to develop a landing system. It all comes down to a simple matter of wavelength. Radio landing systems are based on radio beams, as are many en-route systems. For long-distance navigation, radio beams defined by long wavelengths are suitable. If an aircraft can be kept to within a wavelength or two—that is, say, three hundred meters—a relatively low-frequency signal will do the job. The technology of the 1920s and 1930s was sufficient to generate these frequencies. For landing, the landing system should be about one hundred times more precise. This was way beyond the capability of the electrical art until the war years.

It wasn't as if the industry did not try to develop a landing system using shortwaves. New tools were developed, and progress occurred in short-wavelength navigation and landing systems. In the late 1930s RCA had developed an experimental range station based on the low-frequency AN range but using UHF. Recall that "ultrahigh" in the 1930s was above 30 MHz, which is not "ultra" by modern standards. Frequencies from 30 MHz to 60 MHz are subject to long-distance ionospheric propagation during the peak of the sunspot cycle. Unless the upper part of the "ultrahigh" frequency range were used, this system was doomed to failure.

RCA was also working on an omnirange system that operated in the 90-MHz region. Frequencies this high are not subject to ionospheric propagation at any time and are not far from the modern 108–118-MHz band set aside for VHF navigation aids. RCA was developing transmitting and receiving equipment for use with an important invention of theirs: television. Just as radio broadcasting produced a steady cash flow for the development of electronics during the early days of radio, the potential revenues from television broadcasting would fund VHF technology for other applications.

In 1941 experiments with microwave markers operating at 750 MHz as an aid to landing were carried out at Boston's Logan Airport. If 30 MHz was "ultra," it is not surprising that 750-MHz waves were "microwaves" at the time.

The National Bureau of Standards, which was involved in the development of the earliest radio navigation systems, was also involved in the development of a landing system. A five-hundred-watt transmitter operating at 93.7 MHz transmitted two beams modulated with two tones. A cross-pointer indicator, similar to the modern instrument, was used for pilot guidance. Marker beacons operating on the same 93.7-MHz frequency provided position fixes. The Bureau of Standards ceased work on this system in 1934, but the basic technique was used for the instrument landing system.

At the time the Bureau of Standards ceased development of its landing system, United Airlines, along with Eclipse Aviation, a wholly owned subsidiary of Bendix Radio, worked on a landing system operating at a carrier frequency of 278 kHz. Eclipse is known for its "earth inductor" compass, which Charles Lindberg used in the *Spirit of St. Louis*. The Bendix/Eclipse system was a step in the wrong direction, however. It suffered all of the problems tied to long wavelengths. For a landing system, power lines near the airport were a particularly serious problem. Bendix switched to 93 MHz and used Yagi-Uda antennas, the first navigation system to do so.

In 1938 the Civil Aviation Authority (CAA), along with the Massachusetts Institute of Technology, MIT, jointly worked on a 700-MHz system with superimposed radio beams. The use of 700 MHz was good, but the state of the art in 1938 would not have permitted a price-effective airborne receiver. Landing systems were under development in Europe in the 1930s as well. One promising system using UHF, in modern terms, was developed by Elektrik Lorentz, a German firm, but Germany was too occupied with its war effort to bring the system to production.

By 1942 there were five privately owned landing systems used only for training, three operating CAA systems, and six experimental systems. In addition, the CAA had contracted for the purchase of sixteen additional systems.

The lack of a landing system was not due to insufficient effort but to the lack of a unified effort, which was characteristic of the period. The turbulence of war prevented industrialized countries from working together to develop an effective, worldwide system. The development of shortwaves suitable for precision navigation would open up radio applications in the unused spectrum above 60 MHz. However, shortwaves were the key to radar, which was a jealously guarded secret. The establishment of the CAA in 1938 unified the efforts in the United States. The establishment of ICAO six years later, with the goal of international standardization and cooperation, would be a major move in the development of future navigation aids.

True All-Weather Flying

Every traveler knows there is no such thing as all-weather flying. Extreme weather conditions can, even with the latest technology, ground aircraft. The ILS is certified for three basic categories of bad weather: I, II, and III. In "pilot speak" these are referred to as "CATs": CAT I, CAT II, and CAT III. CAT I is the easiest and is for a ceiling of 200 feet and a runway visual range (RVR) of one-half mile. RVR is the visibility down the runway. CAT II is for a ceiling of 100 feet and an RVR of 1,200 feet. CAT III has three subcategories: CAT III A means a ceiling of 100 feet and an RVR of 700 feet. CAT III B means a ceiling of 50 feet and an RVR of 150 feet. With a ceiling this low, to the crew in a large aircraft such as the B-747, the visibility from the flight deck is essentially zero. The most difficult category is CAT III C, which is zero ceiling and zero RVR—the ultimate.

At the time of this writing all three categories and subcategories are installed and operating throughout the world except CAT III C. Probably the major reason CAT III C is not installed is that there is no provision for taxi guidance. In the zero-zero situation, once an aircraft has landed, it needs guidance to simply taxi to the gate. It is true that all-weather flying may never be possible. Weather conditions such as heavy snow, extreme winds, and ice will not benefit from any type of navigation system. In poor visibility, an aircraft's capabilities are more a function of its electronic equipment than any other characteristic of the design.

Postwar Military Navigation

In the late 1940s the U.S. military set out to design an improved navigation system similar to VOR, called tactical air navigation, or TACAN. It would have been odd to replace the VOR system, which was a huge improvement over previous navigation aids, with a new system before VOR had barely been used. The military was happy with the basic concept of VOR but wanted certain improvements.

The military's main problem with the VOR was that the ground stations were too large and difficult to set up. The military TACAN ground stations were disassembled into cases small enough to be packed up, sent to a temporary airfield, and set up again in a few hours. This problem was easily solved. The basic concept of TACAN is the same as VOR except the wavelength is one-tenth that of the VOR. Therefore, the necessary antennas are roughly one-tenth the length for TACAN. The chosen frequency was around 1,000 MHz. Before the war, when VOR was being developed, the 100-MHz frequency of the VOR was a challenge. After the war, although still not trivial, designing 1,000 MHz transmitters and receivers was becoming commonplace.

Another improvement TACAN provided over VOR was that the former measured the distance from the ground station. If an aircraft knows the distance to a point on the ground, the resulting line of position (LOP) is a circle. All of the points that are a certain distance from a fixed point lie on a circle with its center at the fixed point. Because the TACAN provides a course or a straight line from the location of the ground station, the point at which that line of position intersects the circle is the aircraft's fix. Therefore, TACAN provides an aircraft with course information and a position fix. Because the TACAN ground station was compact, the TACAN navigation aid was installed on ships and other aircraft and was field transportable as well.

Civilian aviation realized that if just the distance part of TACAN were available and the TACAN station were located at the same place as a VOR, the distance portion of TACAN could be used to enhance VOR navigation. The military approved the civilian use of TACAN—not just the distance portion but also the bearing. The distance portion is known as distance-measuring equipment, or DME.

Although microwave technology made enormous advances during the war, an airborne TACAN navigation set was a large, expensive box for most civilian aircraft. Since VOR already provided the bearing information, the DME would permit position measurement. However, the cost of even the DME portion was more than many civilian aircraft owners could afford. This was not the case with large air-transport aircraft. A DME set was well suited for the airliners of the day. With a VOR receiver and a DME set, the air-transport aircraft could stay on a victor airway and know exactly where it was on the airway. This was a significant improvement to airways navigation.

High-Flying Aircraft Fuels a Growing Airline Industry

The air-transport aircraft that were made following the war were a far cry from those flown before the hostilities. The DC-3, which made flying passengers profitable, was obsolete. The postwar, air-transport aircraft would be pressurized, first and foremost. Without pressurization an aircraft is limited to altitudes below ten thousand feet, which relegates the aircraft to flying either under or in the weather. Engines required special treatment to operate at high altitudes, but high-altitude engines had been developed for the long-range bombers, such as the B-17, B-24, and the B-29. The crew on most bombers, however, tolerated high altitudes by wearing oxygen masks, a device that would not endear many passengers. The B-29, however, had three pressurized cabins for most of the crew; only the gunners wore oxygen masks.

Boeing had introduced the world's first pressurized airliner, the B307 Stratoliner, in 1940. The Stratoliner reduced U.S. coast-to-coast flying time from twenty hours in a DC-3 to merely fourteen. The aircraft went into service with Pan American and TWA, but in less than a year the plane was diverted to military service. Douglas's answer to the B307, the four-engine DC-4, was immediately put into military service as the C54 and would not carry commercial passengers until after the war.

The major aircraft manufacturers were quick to apply the lessons and designs of the war. Aircraft coming off the production lines in 1946 and 1947 included the Lockheed L-749 Constellation, the Convair 240, and the Boeing 377 Stratocruiser. The Constellation, or "Connie," cruised at 327 mph with eighty-one passengers and further reduced the coast-to-coast flying time—from fourteen hours in the B307 to just ten. The Stratocruiser was an elegant aircraft with two decks that featured berths for transatlantic flights and a cocktail lounge in the lower deck. With careful observation, one can see the Stratocruiser's resemblance to the B-29.

Although these pressurized, high-altitude, long-distance aircraft were a far cry from the state of air transport before the war, they were nothing compared to what was to be developed in fewer than five years. The first commercial jet, the Comet, manufactured by the British DeHaviland, went into service in May, 1952. The aircraft was way ahead of its time—too far ahead. Pressurized jet aircraft are not like their propeller-driven counterparts. Two aircraft broke up in flight in 1954 due to metal fatigue, a phenomenon not seen before. Although the Comet was then redesigned, it was still a failure. To the traveling public, flying the reengineered Comet was akin to booking passage on the *Titanic II*.

Boeing, however, made its first jet, air-transport prototype in 1954: the B367-80, known as the "Dash 80." The first of the Dash 80 prototypes to be delivered were KC-135 refueling tankers for the air force in 1954. Shortly thereafter, the trendsetting Boeing 707 airliner started deliveries. This was the beginning of the jet age. The new jets were big, high-altitude, long-distance moneymakers. The math is simple. If a B-707 can take a plane full of passengers from New York to Los Angeles in seven hours, it can fill the plane again and make a return trip in the same day. The Boeing Stratoliner required fourteen hours and could not make a round trip in the same day. The Stratocruiser, with a ten-hour flying time, could actually make the round trip but required a rather inconvenient departure time on one side of the flight.

The number of electronics systems in the new jets was greater than ever before in a civilian aircraft. A new industry—"avionics," from the combination of "aviation" and "electronics"—had been born and would become the backbone of what would become the "aerospace industry."

Much of chapter 4 is devoted to the development of radar. After the cessation of hostilities, radar was adapted to the peacetime control of aircraft. Every type of radar needed for air-traffic control had been developed during the war. Long-distance or "microwave early warning" (MEW) radar would be used to control aircraft while en route. Fire-control radar served for the terminal area when aircraft were at lower altitudes and closer to the radar. Even the "identification friend or foe" (IFF) system found an application in air-traffic control.

The Rad Lab formally closed in December, 1945. During its short life of only five years, the lab developed more than one hundred different radar systems—more than half of the radar equipment used by all countries during the war. At the peak of its activity, the lab employed more than four thousand workers and carried out research projects in wave propagation, basic physical science, and communications. On January 1, 1946, the moment the Rad Lab ceased to exist, a transitional laboratory under the sponsorship of the United States Office of Scientific Research and Development opened its doors for business. This transitional laboratory continued microwave-related projects, mostly basic research rather than design and production.

One of the first specific research projects the transitional lab undertook was a weather-radar research project in cooperation with MIT's meteorology department. The project, which started on February 15, 1946, had two main objectives: to understand the physics of the scattering of microwave energy by water drops and ice crystals in the atmosphere, and to use radar to investigate weather.

As mentioned before, aviation and weather have a very close relationship. Doolittle's demonstration of blind flight was a first, important step in all-weather flying. Knowledge of en-route weather permits safe flight planning. The postwar pressurized aircraft allowed for flying above most weather. However, some weather, such as the common thunderstorm, cannot be flown through or over. Even though the modern airliner can eventually fly over most bad weather, it must first fly through the weather. A big danger is encountering thunderstorms embedded in other weather. The capability of locating embedded thunderstorms would be an important tool for the modern, long-range, high-altitude aircraft.

By July 1, 1946, the transitional laboratory was officially established as the Research Laboratory of Electronics at MIT. Seven research staff members, seven technicians, and two administrative assistants directed the weather-radar effort.

The war had produced an enormous amount of surplus equipment; much of it was sold for scrap, and some found its way to the bottom of the ocean or became landfill. Fortunately, some of the surplus equipment was donated to universities and research organizations such as MIT's electronics lab. A ten-centimeter SCR 615 B and a three-centimeter AN/TPS-10 radar set were provided to the electronics lab along with the loan of a specially instrumented B-17, complete with a crew. Because the research projects were under the auspices of the meteorology department, the main thrust was the study of weather as observed by radar. However, the results of these investigations led to the development of both airborne and ground weather radar. Major manufacturers of airborne weather radar for air-transport aircraft were RCA and Bendix, while Westinghouse was a major supplier of ground-based weather radar. Weather radar sets were installed on some of the last propeller-driven aircraft, many of which had to be modified to provide a radio-transparent housing, or "radome," in the nose of the aircraft. The new jets were automatically built with a radome nose. Before the transistor appeared on the scene, weather radar was built with vacuum tubes, which resulted in a very large, heavy, and power-hungry set of equipment. However, the increase in safety the radar provided was well worthwhile.

When jet aircraft began to appear stuffed with avionics, the scheduled airlines realized that standardization of electronic equipment aboard aircraft would be essential to the operation of modern jet aircraft. Standardization had been discussed previously relative to how a system works. The standardization considered by the airlines at this point was for common hardware.

Large aircraft are virtually custom made. Although an airline cannot change the basic design of an airframe, it can specify the type of interior, the paint scheme, and the electronics. The airlines realized that, if all of the avionics a company used were the same or at least interchangeable, fewer spares would be needed. There would be less test equipment, and technicians would need to be competent on fewer systems. The airlines called upon ARINC to draft specifications for avionics in air-transport aircraft. These specifications spelled out the "form, fit, and function" of the equipment. A good example of this type of specification is the common light bulb or fluorescent tube. If a bulb or fluorescent tube burns out in a lamp, it is a simple matter to buy a replacement. Any manufacturer—foreign or domestic—can make this bulb. It has the same shape, operates at the same voltage, and provides the same amount of light. To the airlines, if an avionics box fails on an aircraft, it can simply be removed and replaced by another unit regardless of the manufacturer. In a far-flung operation such as an international air carrier, having a large stock of spares at every airport serviced by the airline would be prohibitively expensive. Interchangeability of spares is an absolute must.

Defining a New Air-Traffic Control System

With the increased volume of air traffic after the war, it was important to have a frequency-use plan for the new VHF frequencies—108 MHz to 132 MHz—assigned for aviation use. Of this range, 108 to 118 MHz was assigned for navigation, while the remainder, 118 to 132 MHz, was assigned to communications. Both the navigation and communications bands were assigned 200 kHz channels, which would allow fifty channels for navigation and seventy for communications. By modern standards this frequency allocation is extremely wasteful, but the level of available technology of the time made it necessary.

Seventy communications channels may sound like a lot, but, because an aircraft at ten thousand feet has a radio line of sight of more than one hundred nautical miles, a large number of aircraft could be within the radio range of a major metropolitan area. One scheme for maximizing the number of communications channels is to use the navigation channels for simultaneous communications and navigation. This is actually possible, as the VOR and localizer were both designed with the capability of simultaneous voice transmission.

For a navigation channel to be used as a part of a communications link, the navigation channel can be used for only the ground-to-air link. The air-to-ground link must use a communications channel. If the navigation frequency is used for the air-to-ground link, aircraft would transmit on the navigation frequency and interfere with the navigation signal. The advantage of using the navigation frequency as a ground-to-air link is that the air-to-air range is twice that of the air-to-ground range. For example, an aircraft at 8,000 feet has an air-to-ground range of approximately 100 nautical miles. The same 8,000-foot altitude produces an air-to-air range of 200 nautical miles. This implies that air-to-air interference covers twice the radius, or four times the area, of air to ground. If aircraft transmit only on a particular frequency, they will not interfere because they do not receive on that same frequency. This technique is called *crossband* because one of the two frequencies is actually in the navigation band. Transmitting and receiving on the same frequency is the more common mode of radio communications, and this is called *half duplex.* Transmitting to an aircraft without two-way capability, the original method of airborne communications, is called *simplex.*

There are some disadvantages of the crossband system. The first and most serious is that the aircraft hears only the ground controller. Thus the aircraft knows only the replies to queries and not the questions. Knowing the position and intent of nearby aircraft is a definite plus, particularly in the terminal area. Second, the complexity of the communications transceiver is greater. Transmit and receive frequencies must be selected separately, and the total number of receive channels is 120 rather than 70. These represent the channels as they were in 1946. For modern technology this is not a problem, even in light of the larger number of channels, but it was a big problem for late-1940s' technology.

In 1946 the CAA, the predecessor to the FAA, requested the formation of the Radio Technical Committee for Aeronautics (RTCA), a special committee to formulate recommendations for the use of the new VHF communications frequencies. The committee members included the Army Air Force, the CAA, the Air Transport Association (ATA), the Aircraft Owners and Pilots Association (AOPA), and members of the Manufacturers Advisory Committee.

The committee made the following recommendations for the use of the new VHF communications frequencies:

1. between en-route aircraft and CAA ground stations: duplex
2. between scheduled air carriers and company stations: simplex
3. between air-traffic controller and aircraft under control: crossband, with rebroadcast of the air-to-ground communications by ground stations
4. between control towers and aircraft in flight: crossband
5. between control towers and aircraft on the airport surface: crossband and simplex

A CAA air traffic control center, ca. 1946. At the time, there was no way to contact the pilots directly, and all communications were via telephone and the control tower. The boards in front of each controller displayed the flight progress strips.

A year after making these recommendations, the RTCA was again asked to consider the simplex/crossband question, except this time for international air travel. Another special committee, SC-34, was formed to make recommendations. RTCA was not formed to be an international organization, but this was one of the first of a number of activities with international repercussions. Eventually RTCA would accept foreign members and thus become international in makeup as well as scope.

Radar: The Backbone of Air-Traffic Control

Air-traffic controllers of the 1930s would have at their disposal a telephone, maybe a radio, and a large map with a supply of wooden "shrimp boats." The boats held the flight information for the aircraft a controller was following and were pushed around on the map as new position information became available. The time between position updates could be quite long, and the situation the shrimp boats portrayed on the map could be quite old—fifteen minutes old or more.

An air traffic control center from 1962. The controllers are now in direct radio contact with aircraft and each controller has a radar screen.

Air-traffic controllers of the 1940s, just after the war, would have the same shrimp boats and map and fewer telephones, but more radios and teletype machines. The scene in the 1950s would include a new and very powerful tool: radar.

Electronic Systems Get Technical Standards

Since the earliest days of aviation, the federal government had imposed standards on aircraft hardware and design. This was because unsafe aircraft jeopardized the lives of not only those on board but of many on the ground as well. With the swell of new systems—VOR, ILS, and VHF communications—what an aircraft could do was defined by the electronics on board. Engines and airframe determined how high, how far, and how fast planes could go, but electronics defined *where* they could go. To fly a federal airway, to fly in or over weather, and to land in controlled airspace—all of these were defined by the electronics systems on board. Certain technical requirements were placed on key components of aircraft, such as the engines and propellers, because failures of these systems in flight could result in disaster. The same was becoming true of the electronics on board. An aircraft making a landing in bad weather cannot lose part of the instrument landing system. Likewise,

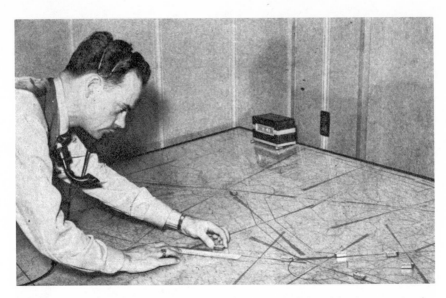

A flight coordinator moves a shrimp boat along an aeronautical chart. The chart contains the beams for the AN range navigation system. Other than the aeronautical chart, the coordinator has only a telephone headset and a clock.

an aircraft that is making an instrument flight on a federal airway and goes off course because of a malfunctioning VOR receiver can find itself out of fuel and lost.

In 1947 the electronics systems that required approval were given a type of certificate similar to those given to aircraft. This was a cumbersome procedure that would have difficulty adapting to the rapidly growing population of electronics systems. The CAA planned to introduce a new program called the "technical standard order," or TSO, intended to regulate airline avionics equipment. This plan would require manufacturers to certify that they have submitted their equipment to tests and that the equipment has passed those tests. Once a TSO has been granted, the equipment may be labeled with the appropriate TSO number. To maintain a TSO, the basic design of the equipment may not be changed without new tests. Minor changes, such as changing the paint on a box or making a small change in a component, may take place without additional testing. A unit that has passed all of the required testing is said to be TSOed.

The RTCA was given the task of formulating the performance requirements and the test procedures for various aviation electronics. Designating the test procedures was an unusual part of a specification. Because many of the systems were new and state of the art, there were no industry-wide, standard test procedures. Therefore, to ensure uniformity of performance, a test procedure was also specified. The caveat "or equivalent" was attached to the test procedure, but it was up to the manufacturer to prove that the alternate test

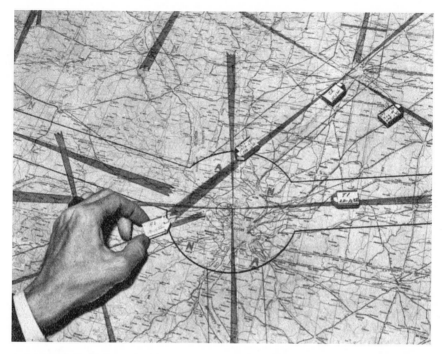

A closeup view of the map used by the coordinator clearly showing the *A* and *N* regions of the AN range and shrimp boats.

procedure was in fact equivalent. This was similar to the problem that the Rad Lab faced. Developing microwave technology was of no value until test equipment and test procedures were developed.

The TSO system remains in effect today in virtually the same form as in 1947. TSOed equipment is required mostly for air-transport aircraft with a few pieces requiring a TSO for all aircraft. Many general aviation (GA) aircraft, although not required to have TSOed equipment, have such equipment installed as a guarantee of quality rather than to satisfy a regulation. Improved technology has made it much easier to pass TSO requirements so that even lower-cost systems can be TSOed.

The Semiconductor Revolution

In the mid-1950s, when the Boeing 367-80 prototype was designed, nearly all electronic equipment was still using vacuum tubes. These tubes had reached their highest level of sophistication during the war. Miniature vacuum tubes no thicker than a pencil and no longer than a few centimeters were used in the miniature equipment installed in aircraft. However, they were vacuum tubes, requiring high voltage, consuming considerable energy, and needing replacement every couple of thousand operating hours.

In 1947 the vacuum tube's replacement, the transistor, was invented at a place not unknown to aviation electronics—Bell Laboratories. In the early 1950s, when Boeing began its $16 million gamble on the Dash 80, no transistorized avionics systems were available. Transistors were becoming commercially available but were not capable of operating at the high frequencies now required of avionics. Transistors were typically found in portable radios, where the frequencies involved are below 2 MHz.

The transistor's potential did not go unnoticed. Nowhere was the transistor considered for designs more than in the ballistic-missile business. The World War II proximity fuze endured an environment that was the closest to the ballistic missile. The proximity fuze was very simple and inexpensive and used only four or five vacuum tubes. The electronic systems required to guide a missile would require much more than a handful of vacuum tubes. These applications would require hundreds of vacuum tubes.

When reliable transistors became commercially available, all airborne applications of electronics started using the new semiconductors. One of the first important applications was in the design of power supplies. Before transistors appeared, converting the DC power supplies found on both aircraft and missiles to the higher voltages required for vacuum tubes involved a variety of electromechanical devices. These devices included motor generators and vibrating switches. As the capabilities of transistors increased, these semiconductors found applications in more circuits, resulting finally in all-transistorized equipment. However, in the early 1960s, before the transistor was fully mature, the integrated circuit (IC) appeared. *Integrated circuits* combined transistors and other circuit elements to create entire electronic circuits. The significant advantage of the integrated circuit is extreme miniaturization.

The integrated circuit was immediately designed into aerospace electronics, primarily missiles. Civilian aviation benefited from the new integrated circuits in smaller, lighter, higher-performance components that consumed less power. Eventually, reliability increased as transistors and integrated circuits became more highly developed.

Further Reading

Heppenheimer, T. A. *Turbulent Skies: The History of Commercial Aviation.* New York: John Wiley and Sons, 1995.

Hurturk, Kivanc N. *Individual Aircraft History of the Boeing 707: Forty Years in Commercial Aviation.* Forest Hills, N.Y.: BUCHair, 1998.

Marriott, Leo. *Eighty Years of Civil Aviation.* Edison, N.J.: Chartwell Books, 1997.

Mellberg, William F. *Famous Airliners: Seventy Years of Aviation and Transport Progress.* Plymouth, Mich.: Plymouth Press, 1995.

Chapter 6

The Space Race
Lessons Learned

Although aircraft did not greatly affect the outcome of World War I, it was clear that aircraft would play an important role in the next war, and they did. A similar situation existed after World War II, but the flying bomb and the V2 rocket appeared too late in the war to make a significant difference. It was obvious to military strategists that guided-missile technology would play a major role in the next war: the cold war.

The German V1, the doodlebug, was not actually a rocket but an air-breathing jet that flew at low altitudes; today it would be called a "cruise missile." As discussed in the last chapter, the advantage of a cruise missile is its ability to fly at a low altitude below radar detection. The V2 was a rocket-powered missile that flew at very high altitudes, out of range of any antiaircraft fire. The V2 flew at Mach 2 and faster, which reduced the early warning time and made it nearly impossible to bring the missile down with conventional antiaircraft artillery. The V2 was capable of virtually leaving the Earth's atmosphere to the point of entering space.

The idea of a flying bomb was not new, particularly to the American military. During World War I, in 1917, experiments were carried out on a flying bomb called the "bug." The device delivered about 180 pounds of explosives and was powered by a four-cylinder, two-stroke gasoline engine. The bug was wheelless and was launched from a catapult. An aneroid altimeter set the cruising altitude, and a turns counter ticked off the engine revolutions. When the counter reached a preset value, the fuel valve was closed, and the bomb glided to its target. The basic concept of the bug was very similar to the German V1.

The bug was a good idea but way ahead of its time. The unsophisticated control systems were mostly mechanical. There was no hope for remotely piloting or controlling the bug from the ground. In 1917 installing wireless equipment on aircraft was daunting, and the idea of installing wireless in a flying bomb was even more so. The range of the bugs was only 100 to 150 nautical miles. Worst of all, the reliability of the devices was very poor. During a demonstration of the bug in 1918, the missile ran amok and headed for the military brass who were assembled, sending them running for cover.

In the years between the wars, only Charles Kettering, chief engineer of

General Motors and involved with the original bug, retained an interest in a flying bomb. In 1939 Kettering contacted General H. "Hap" Arnold, chief of the U.S. Army Air Corps, about the possibility of taking the bug out of mothballs. Arnold appreciated the need to employ technology to modern warfare and showed an interest in reviving the bug. Kettering secured a $250,000 grant from the government for development of the bug. Under Kettering's direction, General Motors designed a new, lightweight engine that produced one horsepower for each pound and a half of weight. The Cessna Aircraft Corporation provided a sleek new airframe. Although the bug was vastly improved from its World War I beginnings, the flying bomb lacked the range and accuracy to launch serious attacks on the German mainland.

Thus the second attempt at a flying bomb became a failure, too. With just Kettering at the helm directing a group of engineers there was hope, certainly in the mind of Hap Arnold, that the flying bomb could be developed. What was lacking in the General Motors flying bomb and the more successful guided missiles of the future were two important developments.

First, although the reciprocating engine producing one horsepower for each pound and a half of weight was a significant achievement, it was no match for the jet or rocket engine. When the engine and propeller are considered together, they weighed seven to ten times more than a jet or rocket engine. Every pound saved from the weight of the engine represents an additional pound of fuel, which increases the missile's range. Furthermore, high-altitude, high-speed flight is more suited to jet or rocket engines than to propeller-driven engines.

Second, the guidance systems were mostly mechanical with some electrical assistance. The flying bomb would not become an effective "guided missile" until the mechanical control system was replaced with sophisticated, electronic guidance systems with remote control. The U.S. military did not recognize this need, and the flying bomb was shelved in late 1941 for the remainder of the war. Research and development at General Motors and other venues would center on conventional, long-range bombers.

The bombing runs over Germany resulted in a very high casualty rate. Later in the war, to reduce casualties without decreasing bombing effectiveness, the U.S. Army Air Force outfitted several worn-out B-17 and B-24 bombers with autopilots and loaded the aircraft with high explosives. The aircraft were piloted from takeoff to their cruising altitude, the autopilot was engaged, and the pilots would parachute to safety. A radio-command system would kill the engines in the bombers, allowing the bombers to glide to the ground and explode. The bombers were not guided by radio but with a mechanical autopilot that held heading and altitude. Radio signals were used only to send the bomber plummeting to Earth. This system proved to be ineffective for a number of reasons. The unmanned aircraft were easy to shoot down. Second, the aircraft could not be directed to a specific target and landed randomly. The practice was that of a vengeance weapon inflicting random casualties on innocent civilians while doing nothing to reduce the enemy's ability to wage war. For these reasons the scheme was quickly abandoned.

This decision to not commit research effort to the flying bomb turned out to be a serious mistake that could have allowed a different outcome to the war. During the fall of Germany, American forces sought out and acquired most of the German research and development hardware and scientific data. In this haul they found both the expected and the unexpected. The plans and models for the V1 and V2 came as no surprise, but there were also plans for a V3. A model of a swept-wing aircraft was found in a high-speed wind tunnel. Probably the biggest surprise of all was plans for a jet-powered helicopter.

After the cessation of hostilities in Europe, General "Hap" Arnold asked the renowned aerodynamicist Theodore von Karman to review the captured scientific documents and hardware. Arnold requested a detailed report that would compare the state of scientific development in the United States to that in Germany. The report, which was aptly titled "Where We Stand," was issued in August, 1945. In the report von Karman made the following predictions:

1. Future aircraft will regularly fly faster than the speed of sound.
2. Future aircraft will just as likely be pilotless as crewed.
3. Electronic controls, along with improved propulsion and aerodynamics, will enable future pilotless aircraft to deliver payloads over distances of thousands of miles.
4. Defense against conventional aircraft will be by automatic, electronic, target-seeking missiles.
5. Vastly improved electronic communications systems will enhance air operations.
6. Improved electronic command, control, and navigation systems will make air operations virtually independent of weather conditions.

In spite of the Allied victory, von Karman criticized the U.S. military of not pursuing scientific research and development. He praised the activities of the MIT Rad Lab and stressed the need for continued development of radar. Von Karman suggested that Rad Lab–style research and development be carried on in other areas such as aerodynamics, propulsion, and electronic guidance. Von Karman, an aeronautical engineer, recognized the important role of electronics. Most of his predictions for future air power were based on the wide application of electronics. In Hap Arnold, von Karman had an eager listener, and, through Arnold, von Karman's ideas and philosophies played a big part in the development of the American aerospace industry.

Eye in the Sky

Unfortunately, the cessation of hostilities in 1945 brought about only a short-lived euphoria. The menace of communism and Soviet domination was a clear and present danger. The Soviet Union, after a period of recovery, would surely develop the future tools of war: missiles, atomic/nuclear weapons, and electronic systems. The United States would put down their arms but not their military intelligence.

One of the very first uses of aircraft was for aerial observation. During World War I, long before effective aerial bombing, aircraft were used to observe troop movements and direct artillery fire. Before airborne communications, it was necessary to return to base before military personnel could analyze and used the gathered information. For the most part, observation aircraft were modified aircraft originally intended for other purposes. Modified P-38s, P-51s, and B-29s were equipped with cameras and surveillance receivers and used for reconnaissance. Near the end of the war, the U.S. Army Air Corps recognized the need for an aircraft designed specifically for reconnaissance. After the war, the now "Air Force" made use of modified B-36s and B-45 and B-50 bombers for high-altitude reconnaissance.

Short wavelength radio signals—signals with wavelengths shorter than about ten meters—propagate in straight lines. VHF and UHF radio frequencies were being used more often than ever before because of their limited propagation. Their use would require an elevated platform for surveillance receivers, a fact that was learned in World War II. The radio horizon is slightly longer than the optical horizon and, just as the optical horizon, depends on the altitude of the observer. The relationship of the radio horizon to the altitude is given by the following formula:

$$d = 1.225 \sqrt{h}$$

where h is the altitude in feet and d is the distance to the horizon in nautical miles. If a ground-based listening post had a 100-foot tower, the radio horizon would be 12.25 nautical miles. The actual horizon would probably be less because a 100-foot tower may not clear all of the trees or buildings in the vicinity of the antenna. If the radio reconnaissance were shifted to an aircraft at ten thousand feet, the radio horizon would be 122.5 nautical miles and would be free of all obstructions except the world's highest mountains. Obviously the higher an aircraft can fly, the greater its radio range and the more signals it can intercept.

Detecting radio transmissions from transmitters with wavelengths longer than ten meters, which usually implies HF, has been done since the 1930s. Although used for detecting transmissions from aircraft, most HF transmissions were from ships. Because of its ability to provide over-the-horizon communications, HF is required for ship communications and for aircraft on sorties well beyond the radio line of sight. Radio direction finding, or DF, is used to determine the angle of arrival of HF signals. This procedure is called HF direction finding, or HF DF, or "huff duff." If two observers with known positions determine the angle of arrival of the same signal, they can determine the location of the signal.

Direction-finding research was carried on after World War I both as a method of navigation and rescue and for intelligence. However, in the 1920s the United States fell behind other nations in DF capability. In 1937 and 1938 the Naval Research Laboratory developed a HF-DF system for locating German submarines and aircraft that was installed in shore stations around the world. In 1940, just ahead of the German occupation, Henri Busignies fled from his Paris home and brought to the United States complete plans for a superior DF system. Although this system did not become operational until 1943, it was adaptable to VHF direction finding. With more communications taking place at VHF, particularly in aircraft, there was a need for a VHF DF system.

Locating the source of a radio transmission using DF techniques is only part of the story; understanding the nature of the intercepted signal is also important. Intercepting communications and decrypting coded messages is a major part of what was to become communications intelligence, or COMINT.

The earliest COMINT activities were from ground-based listening posts, but airborne listening posts were required when more communications took place via VHF and UHF. Airborne receivers with wide frequency capability had been installed in aircraft to detect radar transmissions, as explained in a previous chapter. The same receivers could be used to intercept communications. However, successful COMINT requires that the intercepted signals be decoded and the message understood. Signals intelligence, or SIGINT, would mature from the crude, amateur S-27 receiver to a highly sophisticated science during the cold war.

The decrease in defense spending after World War II precluded the development of a specific reconnaissance aircraft. However, the cold war spurred the air force to consider a high-altitude design specifically for reconnaissance. After a number of proposals from relatively small aircraft manufacturers, an unsolicited proposal from Lockheed was received for a high-altitude, reconnaissance aircraft. Sophisticated observation aircraft came about during the early 1950s during the cold war. This new breed of observation aircraft, often called spy planes, included some modified aircraft but also now included designs that were specifically for reconnaissance purposes. These aircraft were made to fly primarily at extreme altitudes well out of range of antiaircraft artillery.

The Dragon Lady

Lockheed's unsolicited proposal led to one of the best-known aircraft in the world, the "Utility Two," or the "U2 Dragon Lady." This aircraft was a product of Lockheed's "Skunk Works," under the direction of Clarence "Kelly" Johnson. This aircraft was unlike any other reconnaissance aircraft at the time and was designed for the CIA. Beginning in 1955 and extending to 1989, a total of 103 U2s were made. About half of the U2s were based on the original design, but later aircraft had an enormous 103-foot wingspan to facilitate high-altitude flight. The amount of electronics surveillance equipment installed on later U2s caused a serious reduction in performance; thus, the additional wingspan was required to correct this situation. These aircraft were fitted with a huge array of electronic equipment—detectors of all types, not only radio listening equipment, but also all types of scanners, infrared, visual, and radiation detectors, and so on.

The extreme high-altitude capability of a spy plane, in addition to avoiding antiaircraft artillery fire, makes the aircraft less visible to radar. Another method of decreasing the ability of an aircraft to be detected by radar is to reduce the "radar cross section," a technique that is called "stealth" for modern aircraft. However, when the U2 was designed, radar cross section was not considered. High altitude was considered to be a sufficient defense against radar interception. When Soviet radar improved to a point where it could easily detect the U2, the aircraft was covered with absorbent foam. These aircraft were called the "dirty birds." However, the aerodynamic drag of the foam degraded the performance of the aircraft so much, the concept was abandoned.

The U2 had tremendous climb ability as it had to climb to nearly seventy thousand feet. The aircraft was not easy to fly and rather unforgiving. The U2 had a stall speed that was only ten knots lower that the never-exceed speed. This meant the aircraft had to be flown very carefully or by autopilot. It would be humanly impossible to hand fly the U2 for the long periods required by the reconnaissance missions. Therefore, Lockheed provided a very effective autopilot. This requirement of using an electronic control system on an aircraft that is otherwise nearly impossible for a human to fly will appear again in later fighter aircraft.

Spy Planes Get Powerful Eyes

One of the more important roles of spy planes was photographic intelligence, or PHOTINT. The first reconnaissance aircraft were fitted with high-quality cameras using conventional film. These cameras served their purpose and over the years took large numbers of revealing photographs. However, film cameras had serious limitations. First, the pictures could not be analyzed until the aircraft returned and the film was developed. Aircraft such as the U2 were sent on very long missions deep into enemy territory. These missions typically lasted ten or more hours, which meant that nearly a half a day would pass before the first pictures could be evaluated. During this period troops could be redeployed safely out of harm's way. If the images could be recorded electronically, they could be transmitted back to the home base and analyzed before the recon aircraft returned. If additional images were required, these could be obtained before the aircraft returned.

Photographic film reached a high level of sophistication by the mid-twentieth century. It was a process that had been in use for nearly a century and a half. However, electronic imaging held much promise. The first electronic-imaging system was used in the common television camera. By using electronic imaging, pictures could be enhanced in "real time" and images analyzed while an aircraft was over an observation area. The biggest deficiency of PHOTINT was that it would be useful only in clear weather. This was not a new problem. Seeing through weather was the problem that had led to instrument flight. Seeing through weather was also a major reason for the invention of radar. What spy planes needed was a radar system that could produce images of the ground.

Essentially any airborne radar can do what is called a *ground paint*, which produces a crude map of the ground in front of the aircraft. A radar that produces images with resolution approaching the quality of a photograph is the synthetic aperture radar, or SAR. The SAR is a side-to-side-scanning radar that uses not only time-delay measurement to determine range but also Doppler shift from the return echoes to determine relative velocity.

Synthetic Aperture Radar

The ability of cameras or telescopes to observe objects that are very weak or to provide images with high resolution is directly related to the size of the lens or mirror. Every camera buff knows that large-diameter lenses are key to taking pictures in low light or with good clarity. The conventional cameras the early spy planes used were so large that the aircraft had no excess capacity to take other surveillance equipment on board.

Radar imagers have exactly the same problems except that their capabilities are relative to the size of the radar's antenna. To probe a long distance or to produce images of high resolution, radar requires a narrow beam. A narrow beam requires a physically large antenna. The size of the lens, mirror, or antenna of an imaging system is called the *aperture*.

When a photographic camera takes a picture, the shutter opens and admits the light energy of the image. This energy falls on the film or a photo imager, and the entire picture is stored in an instant. The same is true of a conventional radar. A pulse of radio-frequency energy is transmitted, and the radar antenna receives the return signals from the targets. Just as light energy falling on photographic film, all the energy that falls on the radar antenna is captured in an instant.

In synthetic-aperture radar, SAR, the antenna moves across an area, all the while gathering energy from a reflected radar beam. The final image is made from assembling and processing the images made over a period of time. This technique requires a moving platform and extensive signal processing. The gathered data include the Doppler shift of the received signals, which is a very important part of the signal processing. The typical, airborne SAR radar beam looks from side to side. The Doppler shift is positive for signals reflected from ahead of the aircraft and negative for signals reflected from behind. Reflections directly to the side of the aircraft show no Doppler shift. As the beam scans from side to side, the data are stored, including the Doppler shift, and then processed to create a high-resolution, two-dimensional, ground map. Because quite a bit of data processing is required, the SAR requires a complex computer.

A big problem associated with the SAR is imaging moving targets. This is because the Doppler shift of a received signal is a function of the relative velocity of two objects. The added velocity of the moving target tends to obscure its image. Many improvements have been made to the SAR to handle all types of problems. One solution to the moving-target problem is to use conventional radar to locate these targets and place them on a map drawn by a SAR. This is the technique behind the Joint-Surveillance, Target-Attack, Radar System, or J STARS.

One of the major advantages of SAR imaging is its ability to generate images through cloud cover. Similarly, depending on the wavelength of the SAR, a high-performance SAR can penetrate vegetation such as tree canopies. The modern SAR can create images with a resolution of better than one foot from high-altitude spy aircraft.

The U2 was designed at a time when reconnaissance aircraft had a camera and a couple of radio receivers. To accommodate all of the new electronic devices, the U2 used "pods." These were aerodynamic enclosures that hung under the wing of the aircraft much like a bomb under a fighter or bomber. In fact, many fighters and bombers could carry either ordnance or reconnaissance pods hanging from the bomb racks. The U2, however, had so many electronics devices in the pods and in the fuselage that a new model, the U2R, was designed with larger engines. The amount of electronics on later reconnaissance aircraft was staggering. Transistors and integrated-circuit technology had reduced the size of electronics systems, but the systems grew in capability, and the electronics filled the available space.

Passive Imaging

A photographic camera is a passive device in that it receives light reflected or radiated from an object. A radar system is more like a flash camera in that it provides its own illumination. Both systems have advantages and disadvantages. First, a radar imager can take pictures at night, but the radar can be detected and jammed because it radiates energy.

Infrared imaging is a passive system that uses the longer wavelengths of light. Unlike visible light, many objects such as human bodies, vehicles, buildings, and roads give off infrared energy. Also, infrared energy can penetrate a certain amount of cloud cover. Infrared imaging systems can be used without radiating energy and therefore have a low probability of intercept, or POI. Probably the best-known infrared imaging system is called the forward-looking infrared imaging system, or FLIR. These systems are not only used for military aircraft but have also found widespread application in police work and search and rescue. Airborne FLIR systems are extremely powerful tools.

Even at night there is some light energy from the stars, moon, and synthetic sources. As good as FLIR systems are, they cannot be used with objects that do not give off significant infrared energy. This includes unheated buildings, equipment that has not been operated for some time, or equipment that has been deliberately insulated. Light-amplification systems are used for enhanced night vision. One very common application of this is in night-vision goggles, or NVG. This device is used by ground troops as well as aircrews. NVG is a light-amplification system that can provide magnifications of 30,000–50,000x. This means that a moonlit night is as bright as day even for a small sliver of moon. The more sensitive NVG can view scenes with just starlight available, even in partial overcast.

For aircraft use, NVG are particularly suited for nighttime helicopter operations. One limitation of NVG is the limited field of view. The human eye has a field of view of about 200 degrees, while NVG vary but have a field of view of only 20 to 30 degrees. This narrow field means the wearer must use more head movement to scan out-the-window scenery or the instrument panel. For this reason the typical helicopter NVG operation generally has only one crewmember wearing NVG. Crewmembers without NVG scan the instruments and do wide scans out the window.

The most significant problem with the use of NVG is overload from light sources. Out-the-window sources such as vehicle headlights, flares, and so on are problems. However, sources within the cockpit are also troublesome. If the illumination of the instruments were reduced so that it did not cause overload of the NVG, then the crew members who were not wearing NVG would not be able to see the instruments. There is a rather clever solution to this problem. The NVG tend to be more sensitive to the longer wavelengths, or the redder colors. The eye is more sensitive to the green part of the spectrum, or midwavelengths of the visible spectrum. Therefore the instrument lighting for the aircraft is near the blue-green part of the spectrum, where the eye is reasonably sensitive, and the NVG are provided with a blue-blocking filter. This filter does not materially affect the sensitivity of the NVG but ensures additional protection from the instrument-panel lighting.

Even for some civilian applications, such as law enforcement and search and rescue, NVG would provide a safety enhancement. However, for air transport, both private and commercial, NVG would provide few benefits and cause many problems. The NVG would improve pilot vision, which is an obvious benefit. However, the majority of the time, air-transport aircraft are flying at high altitudes en route, and there is no history of collisions due to poor visibility during this phase of flight. This is because en-route aircraft follow a flight plan and are constantly tracked by air-traffic radar. Most accidents that are not due to mechanical failure occur during poor weather. This lack of visibility is not due to lack of light but rather to fog. NVG cannot improve visibility in clouds or fog.

The environment in which air-transport aircraft operate has a plethora of light sources, and the problems with overload would be great. The cockpit would have to be well sealed from the cabin. The aircraft's own external lights, such as beacons, marker lights, and landing lights, would have to be very well shielded from the NVG. Because of so many possible problems and very few benefits, very little use of light amplification systems has been made in civilian aircraft, with the exception of law enforcement and search and rescue.

A scheme of enhanced vision that does not use light amplification has been contemplated and experimented with for both military and civilian aircraft. This system, called *synthetic vision,* uses wavelengths longer than visible light, such as infrared and very short microwaves, called *millimeter waves.* These wavelengths penetrate clouds and fog and can be used to synthesize a visual picture. In a simplistic view, this system takes information from FLIR and from radar systems and creates a visual picture for the crew. Work continues at the time of this writing, and the results look promising.

Not all reconnaissance is from high altitudes. Police helicopters, for instance, are low-altitude observation aircraft. The high-altitude spy planes were involved in SIGINT, but many applications, for example, submarine and coastal smuggling detection can be satisfied with lower-altitude aircraft. Also, SIGINT not involving overflights can be done with low-altitude aircraft.

The U2 had many problems, one of which was that it was very difficult to fly. The U2 carried only one crew member, and in spite of the amount of electronics stuffed into the aircraft and on pods, its payload was limited. Aircraft such as those of the venerable C-130 series can carry large crews and enormous amounts of surveillance equipment and can be fitted with a vast array of antennas. Once believed to be untouchable, the high-altitude spy planes are easily downed by modern missiles. Satellites have to date not been "killed" by hostile forces. This is not to imply that they can't be, however. The problem with aircraft is that they are too close to the ground, so they are vulnerable to missile attack, whereas the satellites are too far from the ground, making it more difficult to observe activities on the ground. However, extreme advances have been made in remote imaging and data processing, and modern satellites can gather reconnaissance data as well as a spy plane. The best platform for an SAR is a low, Earth-orbiting (LEO) satellite, which is one of the reasons the SR-71 spy plane was retired and there are no plans to make new U2s. The chapter on high-altitude spy planes is now closed.

Entry into Space

We are discussing the role of electronics in the evolution of flight. We could ask, "is space travel actually flight?" The words of a popular song go, "fly me to the moon," but do we actually "fly" to the moon? We make the trip in a vehicle that has no wings. Once we leave the Earth's atmosphere, we coast through the emptiness of space during the vast majority of the trip. There has been the idea of an aircraft that can fly to the limits of the atmosphere and then simply rocket into space. The space shuttle achieves half of this goal in that it is launched into space on a rocket and flies home. Whether space travel is really flight is not an important issue in the role of electronics in the evolution of flight. Whether we are talking about "flight" or "space," the cast is the same. The major players in the space race include Lockheed, Boeing, McDonnell Aircraft, Douglas Aircraft, Grumman, and the Martin Corporation. All of these names had a long history of manufacturing aircraft and became very large companies making military aircraft during the war.

The first ventures into space were with unmanned vehicles. The German V2 ballistic missile was one of the very first vehicles to travel in space. A ballistic missile consists of three major parts: the payload, the propulsion system, and the guidance system. It might seem strange that airplane builders with no expertise in these three components would be the major players in missile development. First, the payload is a warhead, which would be supplied by the military. Second, aircraft manufacturers developed propulsion systems, but they were reciprocating engines and propellers and, later on, jets. None of these is of any value in space. Finally, aircraft were piloted and had an entirely different guidance system. Although there were some automatic flight controls, such as a simple autopilot, aircraft manufacturers had no experience in fully automatic guidance systems during the early days of space exploration. What the large aircraft manufacturers had to offer was the ability to make the shell of the missile, which was essentially a large fuselage, and then install the other components. The aircraft manufacturers did not know it, but they had become *systems integrators.*

Only the very early aircraft manufacturers such as Curtiss and Wright made both engines and airframes. An interesting twist of fate befell the Curtiss and Wright companies. In 1929, after the departure of their namesakes, these two companies fell on hard times and decided to merge. An earlier chapter discusses how the Wright brothers and Curtiss were adversaries in the very early days of aviation and were involved in nasty patent litigation. Curtiss was an engine builder, while the Wrights built a superior airframe. The irony of the Curtiss-Wright merger is that the Curtiss division of the company made airframes while the Wright division made engines.

Probably the last aircraft to use engines made by the airframe manufacturer was the Ford Tri-Motor, which lasted until 1932. By the time the manufacturers of air-transport aircraft, such as Boeing, Douglas, and Lockheed, started making serious passenger aircraft, designs centered around a power plant from an engine manufacturer.

One part of an aircraft that the airframe manufacturers never made in the early days of aviation was the radio equipment. Airframe companies understood the concepts of reciprocating-engine manufacture, such as casting, milling, boring, drilling, and other machining operations, as they performed many of those operations themselves. Engines were mechanical; manufacturers could lay their hands on engines and understand them. Radio, however, was totally foreign to the airframe companies. To be fair, radio technology was foreign to most corporations in the 1920s. The strange components in a radio and the invisible power of the device just weren't like an engine.

Cessna, the largest manufacturer of light aircraft, acquired Aircraft Radio Corporation in the early 1950s and for thirty years manufactured their own avionics. This turned out not to be a good idea because most airframe manufacturers did not make their own avionics; airframe manufacturers just do not understand electronics. More recently, Boeing, one of the few remaining large airplane companies in the United States, has started to manufacture various electronic systems for their aircraft. Boeing is being very cautious to make only the less critical avionics, where the volume is high and sophistication is low, such as passenger-service electronics.

During World War II, the concept of a systems integrator became established. Even though Boeing, for example, made the fuselage, another company made the engines; another, the wheels, tires, and ailerons; and yet another, the elevators and rudders. It was necessary to employ as many capable factories as possible to turn out the required volume of aircraft for the war effort.

World War II avionics were made by a number of companies, most of which were not avionics companies but were building to print. Small aircraft companies such as Piper and Cessna were making ailerons, rudders, and elevators for the larger Boeing aircraft. Boeing put it all together and ensured that all the pieces were compatible. Because an aircraft is a large system, Boeing's role was that of systems integrator. As aircraft became more electronic, most of the systems to be integrated were electronic. This forced the major airframe manufacturers to establish electronics departments in their companies to play an important role in the design of aircraft.

When ballistic missiles became a prominent weapon of war, no one company had in-depth expertise in all three missiles components: propulsion, guidance, and payload. When the U.S. Defense Department began letting contracts for the design and construction of missiles, it sought a "prime contractor," or systems integrator. The prime contractor would be responsible for selecting the subcontractors to provide the individual systems for the missile and then assemble the subsystems into a fully operational system. The concept extended beyond just hardware. The prime contractor provided installation, maintenance, and training. The role of systems integrator that had been learned during the war was carried over into the production of modern air-transport aircraft.

Most of the interfacing of a missile's systems was through electrical signals. The guidance system provides electric signals that direct the thrust of the rockets and then release the payload at the right moment. The airframe manufacturers were writing electrical specifications, providing the interconnecting wires, and testing the electrical systems. The major airframe manufacturers found they were hiring more electrical engineers than ever before.

After World War II, the United States confiscated several V2 rockets from the Peenemunde rocket research site and brought them to the United States. The United States brought not only the rockets but also the primary researchers, Werner von Braun and 132 members of his team. The V2 stood 46 feet high and could carry a warhead at 3,600 miles per hour at altitudes of over 300,000 feet.

A number of companies were involved in the design and manufacture of rocket motors. One of the first was Reaction Motors, Incorporated, a New Jersey company that was formed in 1945 for the sole purpose of making rocket motors. Reaction Motors is best known for making the rocket engine for the Bell X-1, the first supersonic aircraft. Other rocket manufacturers also appeared, such as General Electric and the Rocketdyne division of North American Aviation.

One of the major suppliers of guidance systems was the Sperry Gyroscope Corporation. Sperry's dominance was nearly a certainty because of the nature of the guidance system. Ballistic missiles must have an autonomous guidance system for a variety of reasons. First, if a missile is to be remotely controlled, the controlling signals must be via a radio link. This option presents two very serious problems. First, enemy forces can jam the radio signal, causing the missile to go astray. Second, the missile's range is well beyond radio's line of sight. The missile is controlled via an inertial guidance system, which relied heavily on gyroscopes in the early development.

Aside from the gyros, the remainder of the guidance system in the V2 was primarily mechanical. A rule of electronics says that "any task that is performed mechanically will be greatly improved if it can be done electronically." There are a number of well-known examples of this. Compare a mechanical adding machine to a handheld calculator. Or think about what happened to the typewriter. In 1946 MIT began designing an aircraft simulator using a mostly mechanical computer. After nearly four years, the project was over budget and far from complete. Project engineers decided the electromechanical computer was simply too slow and too crude for the task, so they abandoned the project. Compare this with the modern aircraft simulator, where, with the exception of the motion controls, everything is electronic. Its level of simulation far transcends what even the most complicated mechanical device could provide. Simulating the motion of an aircraft and controlling the motion of a real aircraft takes similar computers. It did not take much prodding to convince the engineers who were designing the first missile-control systems that a mechanical computer would not work.

Every possible function of a missile-guidance system was converted to electronics, and the improvements were substantial. This move toward electronics brought a different group of manufacturers to the aerospace industry, including Philco, the Radio Corporation of America, Westinghouse, General Electric, Raytheon, and Magnavox. Familiar names? These are some of the pioneers of the Golden Age of Radio, which was now a mature twenty-five years old. Before the war these companies' names were attached to the wooden cabinets of home radio receivers. Later these names would be on the metal identification tags of military electronics. These companies had designed and built millions of electronic assemblies for the war effort and are now the proud owners of fat bank accounts. They also had the engineering talent needed for the new paradigm.

In 1949 the Northrop Aircraft Corporation developed and shipped a stored-program digital computer. This computer, the binary automatic computer, BINAC, was one of the first airborne digital computers and was used on the long-range SNARK missile. The SNARK, however, was a cruise, or low-altitude, missile, and the rigors of space or near space were not required of the vacuum-tube computer.

The decade after World War II was a period of continued advancement in electronics. Invented in 1947, the transistor was valued for its small size, ruggedness, and low power consumption, all of which made it perfect for missiles. The Texas Instrument Company entered the missile industry with electronic systems using transistors. Later, Texas Instruments would become a leader in the development of integrated circuits, which they used in their missile systems.

Most of the early electronic guidance systems of the 1950s used analog circuits. The digital revolution, which was beginning to invade most electronics, brought in another new player, one that would have a profound effect on all electronics. That player was the Federal Systems division of IBM. Other computer manufacturers such as Univac and Litton soon followed. By the late-1960s, missiles and other airborne weapons had achieved a high level of sophistication. Digital computers were used in many airborne systems, integrated circuits were rapidly replacing transistors, and aircraft and missiles were filled with electronic systems.

Missiles are unmanned vehicles that are capable of entering the fringes of space. The urge for manned space exploration was strong in the 1950s. Piloted vehicles had reached the edge of space several times. The "X" series of rocket planes achieved an altitude of 354,000 feet with the X-15. On October 4, 1957, the Soviet Union launched the first orbiting satellite, dubbed Sputnik, into Earth's orbit. The small, 184-pound spacecraft transmitted a warbling, Doppler-shifted signal to Earth. No special equipment was required to receive Sputnik, and even radio amateurs were able to hear the signals from space. The launch of Sputnik engendered fear in the minds of many, however. Visions of Soviets spying from space alarmed the populations of Western nations. However, only five months later, the American Vanguard satellite was launched into Earth orbit, and the space race was on.

Although Sputnik was the first satellite ever launched, Vanguard was superior. Sputnik lasted only a few days until its batteries went dead. Vanguard, on the other hand, had a solar array that recharged the satellite's batteries. Even more fascinating, in 1961 the first amateur-built satellite, OSCAR-1, or "orbiting satellite carrying amateur radio," was launched and was only the first in a series of more than thirty amateur satellites. Amazingly, with volunteers and a free ride into space, OSCAR-1 cost a mere twenty-six dollars to build. The space race gave birth to a new industry: the aerospace industry.

General Eisenhower attributed U.S. military might to the joint efforts of both military and industrial power. He accurately predicted that military strength of the future would result from the efforts of what he called the "military industrial complex." Eisenhower also warned of the dangers of this powerful alliance, however. The industrial part of this two-headed weapon, for the most part, was the aerospace industry. Von Karman, in his report to Vannevar Bush, who ultimately reported to President Truman, Eisenhower's predecessor, predicted that the capability of future aircraft would depend on electronics. Like Eisenhower, Von Karman accurately predicted the future, as much of the aerospace industry utilized electronic equipment.

All different types of technological advances came from the aerospace industry, financed by a high level of defense spending that was fueled by fears of the cold war. Enormous advances in technology took place, such as basic physics for nuclear weapons, space technology, and information technology. The greatest advances were in electronics, more than in any other discipline.

At the time of Sputnik's launch, the vast majority of electronics still used vacuum tubes, as had been the case from the very first days of radio broadcasting. The transistor was a postwar invention, and its use was slowly finding acceptance, but even a decade after the invention, the vacuum tube still reigned supreme. The requirements for modern missiles, aircraft, and space vehicles virtually precluded the vacuum tube, however. The vacuum tubes that were developed for use in the proximity fuze were comparatively expensive, but that fuze used only a handful of tubes. More complicated systems used hundreds of tubes, and the expenses skyrocketed. Vacuum tubes require high voltage, which required large, inefficient power supplies. Many missiles and aircraft still used vacuum tubes, but it was clear that the transistor would be applied to modern aviation.

Enormous amounts of government funding went to improving and applying solid-state devices to aircraft and missiles, and, of course, these improvements were quickly applied to all electronics. One area in which vacuum tubes were superior to semiconductors was their ability to resist the effects of radiation from a nuclear detonation. There are several types of radiation, including gamma rays, X-rays, subatomic particles, and high-level electromagnetic fields. However, modified semiconductors and other methods were implemented to overcome radiation. Such systems were said to be "hardened." The cost of hardened electronics was so prohibitive that only select aircraft had hardened systems. As a result, the only aircraft to be fully hardened was the original B-1 bomber.

Space electronics are also subject to radiation. Unlike nuclear detonation, where the radiation is intense and swift, space radiation accumulates slowly over years, requiring spacecraft to have hardened electronics. Modern design techniques produce a satellite design life that exceeds fifteen years.

Navigating without Radio Signals

The bulk of research and development of navigation systems depended on radio beams. Some very capable systems were developed and permitted all-weather global navigation. However, radio signals can be vulnerable to a number of disturbances. First, in the military arena, there is always the threat of interception and jamming. Also, natural disruptions (such as solar flares) of radio signals occur. Navigation systems that are independent of radio signals are desirable for military aircraft and as a second navigation system for civilian aircraft. As every student of science learns in basic physics, if the acceleration vector is integrated, that is, mathematically, the velocity vector is obtained. A second integration produces the position vector, or just plain "position" in navigation terms. A system that senses the acceleration of an aircraft and derives its velocity and position is an *inertial navigation system.*

Inertial navigation was used not only on missiles but on aircraft as well. An inertial navigation system can determine in three dimensions how far an aircraft has traveled. Therefore, if an inertial navigation system is "told" where it is at the beginning of a flight, it can then determine its later position relative to the initial point. If the initialization includes the latitude, longitude, and altitude at the beginning of the flight, the system can determine latitude, longitude, and altitude anywhere along the flight path. The inertial navigation system can also provide heading and attitude information.

It is important to realize that a missile's initial position must be known to successfully navigate to a target. Digital communications links constantly inform submarine missiles of their position. Each missile is plugged into the "ship's inertial navigation system," or SINS, and is constantly updated. The same is true of airborne missiles. Missiles and other weapons, while hanging on their racks under an aircraft, are constantly communicating with the aircraft's systems and are informed of their current position and the intended target. This intercommunication is called "stores management" and is responsible for arming, releasing, and guiding weapons. Instead of inertial navigation systems, modern surface ships and aircraft keep track of their position using the satellite-based, global positioning system, or GPS.

The first inertial navigation systems required what is called a "stable platform" on which the accelerometers were mounted. Using gyros in an electronic system to stabilize them, these platforms would drive electric motors and keep the platform absolutely horizontal. Every gyro drifts, so the stabilized platform would have to be reoriented periodically. The better the gyro, the less the platform would drift. Enormous amounts of research went into improving gyros, but as long as the gyro was a mechanical device, drift would persist.

Another problem arose in fighter aircraft. The violent maneuvers of aircraft tended to exceed the capability of the platform to right itself, so the platform would "tumble." Once the platform tumbled, the system had to be reinitialized. It was actually easier to make an inertial guidance system for a missile than for a fighter aircraft. Although a missile travels at extremely high velocities, it does not normally make violent maneuvers that would tumble a platform.

The ring-laser gyro and later the fiber-optic gyro brought gyro performance to new levels. The laser and fiber-optic gyro are somewhat mechanical devices and can still be thrown off by mechanical changes, but they have no moving parts. Even with improved gyros, the stabilized platform was a weak link in the inertial navigation system. As long as the system included a platform, there would be mechanical motion and therefore a weak link. A new concept called the "strapdown" inertial navigation system was invented that required no stabilized platform because it was strapped down to the airframe.

The strapdown inertial navigation system requires a large increase in computing power to process the accelerometer output signals and provide the navigation information. Fortunately, advanced microprocessors were available at the time, and the laser or fiber-optic gyro strapdown inertial navigation system is the standard choice for all aircraft, both military and civilian. The strapdown system has no platform, gyros, or motors; the system is mostly a computer. The early designers of missile guidance systems have essentially met their goal of removing all mechanics from the guidance system.

Global Positioning System

Omega and the LORANs, as were discussed previously, provided aircraft, ships at sea, and other vehicles with latitude and longitude fixes. Unlike other navigation systems, LORAN and Omega provided position rather than course information relative to a point. To provide navigation data to the crew, a selected destination point called a *waypoint* is entered into a computer, which does the geometry calculations to provide steering information to the waypoint. In addition to the steering information, the ground speed and time to go are also calculated. A waypoint is not necessarily the final destination. Like with the VOR, where an aircraft flies from one VOR to the next, these waypoints can be set at specific locations where a visual fix may be made. Waypoints may also be VORs, which would make it necessary to fly the victor airways. Although the navigation is by Omega or LORAN, the waypoints are VORs.

Any long-range, radio-based system is vulnerable to propagation anomalies. When a signal has traveled a long distance, such as with LORAN-C and Omega, the anomalies can be significant. Many of these are caused by weather, primarily noise from lightning. Omega was a worldwide system but had an accuracy of only a nautical mile or so. LORAN-C accuracy is significantly better, but coverage is not worldwide. In 1973 the military, along with civilian contractors, began discussions to develop a satellite-based navigation system that would provide worldwide, all-weather navigation fixes that could be used for all vehicles, including "manpacks." The system was to be called Navstar and would be based on a constellation of orbiting satellites.

Satellite navigation was not new in 1973. In the 1960s a navy navigation satellite system called Transit was launched. The Johns Hopkins Applied Physics Laboratory did most of the scientific work for Transit. The Transit satellite system determined position by measuring the Doppler shift of a carrier. A *carrier* is a simple continuous wave, uninterrupted even by Morse code characters. The concept was simple: If a satellite in orbit passed directly overhead, the Doppler shift would be toward a higher frequency when the satellite rose on the horizon and toward a lower frequency when the satellite set. When the satellite was directly overhead, the Doppler shift would be zero. If the satellite did not pass directly overhead, the zero Doppler shift would occur at the midpoint between the satellite's rising and setting. From this, a line of position, or LOP, could be determined, assuming that the orbit of the satellite was known. If a second satellite was available, a second LOP could be calculated; the user's position is where the two LOPs intersect.

In the Transit system, two satellites would seldom be visible simultaneously. Therefore, the two LOPs would be determined at two different times, sometimes separated by hours. Thus the first LOP had to be updated by dead reckoning using the normal compass and speed measurement and the new, updated LOP with the second, satellite-measured LOP. The Transit system provided worldwide coverage, but because the number of satellites was small, fixes could be obtained only every few hours. This was fine for ships at sea, which move at relatively low velocities, but this system was virtually worthless for aircraft. However, it was an exercise in satellite navigation from which many valuable lessons were learned.

Some of the more important advancements that came from the Transit program were the computer algorithms for accurate, satellite-orbit prediction. For a satellite navigation system to be successful, it is necessary to be able to predict the satellite's position very accurately. Simple models based on a spherical Earth are not sufficiently accurate for the actual situation of an *oblate spheroid* (that is, a sphere that is slightly compressed at the north and south poles).

In 1972 another navy satellite system that used precision clocks was launched. This system, called Timation, was a very important experiment for the future Navstar system. The Timation system distributed accurate time of day to users and was never intended to be a navigation satellite system. The original satellites were fitted with very stable quartz clocks, which were later replaced by atomic clocks.

An air force satellite system called "Project 621" was the original GPS program. Because a satellite navigation system would benefit all branches of the military, Project 621 became a joint military program and was renamed "Navstar." Thus Navstar became one of the first, modern, joint military projects. The Navstar system required the use of very accurate spaceborne clocks. Therefore, the placement of the atomic clocks in space was an important part of the Navstar/GPS navigation system.

Navstar was to be an all-weather, fully global navigation system that would provide an accuracy of better than a meter—everywhere and at any time. It is important to note that this accuracy is relative to the full capacity of the system, or its "precision position service" (PPS). This "military-only" capacity of the system is not available to all users. GPS is also a three-dimensional system in that it provides not only latitude and longitude but also altitude. It is important to understand what aircraft can do with the Navstar system that could not be done before. First, one-meter accuracy is better than any navigation system in operation at the time. A good CAT III ILS can provide this level of accuracy just about at the point of touchdown, but the Navstar system can potentially provide this level of accuracy anywhere in the world at any time. This implies that Navstar could be used for the en-route, approach, and landing phases of flight.

The first satellites launched were called "navigation technology satellites," or NTS. These were an extension of the Timation satellite system and were for the purpose of exploring and learning. The first satellite, NTS-1, was launched on July 14, 1974, and was also the first satellite to carry a rubidium atomic clock into space. The second and last NTS satellite carried a cesium atomic clock and contained a GPS computer. Several of the components of the later GPS satellites were in NTS-2.

The next group includes the navigational development satellites, or the "block I" satellites. These were complete GPS satellites. Only 11 of the 12 satellites planned for block 1 achieved orbit because of the failure of a refurbished Atlas-F booster at Vandenberg Air Force Base. All of the remaining satellites achieved their initial operating capability, or IOC. Although the satellites were designed for only a three-year life, several lasted for more than ten years. The last block I satellite was launched in October, 1985.

The block II and block IIA satellites, or the operational satellites, were first launched in February, 1989, and a total of nine satellites were launched in block II. The launch of the block IIA satellites brought the GPS system to full operational capability, FOC, by the end of 1994. The final block of satellites, the block IIR, or "replacement," satellites are being launched at the time of this writing.

The GPS system is called a one-way ranging system. Distance-measuring equipment (DME), part of TACAN, and radar are two-way ranging systems in that they require an interrogation and a reply, or a two-way communication. In the case of radar, an illumination pulse is transmitted and backscatter is received. The time delay is measured, and from that, lines of position can be determined as discussed in earlier sections. The GPS satellite transmits the time of day as a part of what is called the *navigation message.* This time is compared to a clock in the user equipment. By subtracting the time of day at the user from the time the signal was transmitted from the satellite, the enroute time of the signal is known. From that time the distance to the satellite is calculated.

It is important to understand why extremely accurate clocks are required of the GPS system. Radio waves travel at the same speed as light—about one meter in three nanoseconds. A nanosecond is 1×10^{-9} seconds, or one thousandth of a microsecond, which is a very short time indeed. In order to determine a line of position to within one meter, the time-of-day clock must be accurate to three nanoseconds. A clock capable of such accuracy is no ordinary clock. It isn't an ordinary atomic clock, either. Sending such a precision device into space was not going to be an easy trick. An atomic clock that could withstand the extreme acceleration of the launch and then the rigors of space for ten to fifteen years required a very special and expensive atomic clock.

The GPS system consists of twenty-four satellites in six orbital planes of four satellites each. The original Navstar system had eight satellites in three orbits. Twenty-four expensive atomic clocks for potentially millions of users was still an economically viable situation. However, if the system was to be affordable, the users could not be required to have an atomic clock in their system. Therefore, the original Navstar system used the lessons learned from the time-dissemination, Timation satellite system. The GPS was designed to provide not only precision navigation but also precise time dissemination. The GPS would transmit its accurate time to the user, who would then use that time to determine position.

GPS and its predecessors are military systems, but part of the GPS system is available for civilian use. The GPS has two basic accuracy levels: the standard precision service, or SPS, and the precision position service, or PPS. The signals from space are in a format called *spread spectrum*. This is a technique of deliberately broadening a signal over a large range of frequencies by using randomlike codes called *pseudo random noise*. True noise is called a nondeterministic signal, meaning that it cannot be predicted. Pseudo random noise is a signal that has the characteristics of noise but can be predicted from a mathematical equation. In order to receive these signals, one must know the mathematical equation that generates the pseudo random noise.

If the use of the GPS system were to be denied to all users, the codes would not be made public knowledge. In the GPS system, the SPS is based on public codes, and the PPS was not made known to the public. In spite of its supposed secrecy, the precision code was cracked. However, the PPS is further encrypted, using codes that are changed weekly to prevent further hacking. However, even the accuracy of the SPS, if used by hostile forces, could be detrimental to the security of the United States. Therefore, the GPS standard positioning service can be deliberately degraded from the control stations through a technique called *selective availability*, or SA. Selective availability was set to achieve an accuracy of about one hundred meters; the actual value has been a military secret for most of the life of the GPS system. The irony is that selective availability was turned off at least twice, both times during military actions. Once was during Desert Storm, and the second time, during the 1999 NATO operations in Kosovo.

There are methods of defeating selective availability, primarily with a technique called differential GPS, which we will discuss later. In addition, at an accuracy of one hundred meters with selective availability operational, the GPS system is better than most other navigation systems and could be used for hostile actions in spite of selective availability. Since the GPS system was gaining more customers in the aviation world, selective availability could cause potential problems for these new users and possibly jeopardize personal safety. Selective availability did not ensure military security and was actually detrimental to civilian users. Therefore, on May 2, 2000, selective availability was turned off, and any need to ensure the security of the United States would be handled by jamming the GPS signal in selective locations in such a way that military personnel could still use it.

The use of GPS by aircraft has many far-reaching applications. First, the system has sufficient accuracy to allow its use for en-route and terminal applications anywhere in the world. This means that airports that do not have an instrument landing system because of financial or physical reasons can now have instrument-landing capability. Thus, any airport anywhere in the world can be an instrument field. Even in undeveloped countries, the smallest airport can have instrument-landing capabilities.

Because aircraft using GPS know their positions with great accuracy, they can be separated by less distance anywhere in the world. Recall that the Omega system, which was used for transoceanic navigation, permitted aircraft separation of no less than sixty nautical miles because of Omega's one- or two-nautical-mile accuracy. Transoceanic separation using GPS can be reduced to a few nautical miles—or about the same distance that is used for en-route separation on victor airways when aircraft are under radar surveillance.

These reduced-separation minimums permit several very positive improvements in transoceanic travel. First, more aircraft can safely travel transoceanic routes. More important, aircraft can select more desirable routes, such as more direct for fuel and time economy or improved weather conditions. With wide separations, aircraft have fewer choices.

The GPS system is a three-dimensional system; therefore, the possibility of providing a three-dimensional landing system exists. Unfortunately, the least accurate of the three axes (latitude, longitude, and altitude) is altitude, which requires the greatest accuracy for instrument landings. The greatest accuracy from the GPS system is available from the precision position service or military mode.

The precision position service involves more than just decoding the secret military code. A number of phenomena impair the accuracy of GPS, one of which is the slowing of radio waves through the *ionosphere,* which is one of the higher levels of the atmosphere. The ionosphere is constantly changing, not only from hour to hour but also over an eleven-year period with the sunspot cycle. Solar flares can cause extreme changes in the ionization level of the ionosphere. In a word, the ionosphere is unpredictable. Because of this, it is necessary to measure the delay through the ionosphere rather than calculate it. Fortunately, the delay is a function of the frequency of the radio waves passing through it. If two different frequencies are used for the transmission of information from the satellites, the absolute delay through the ionosphere can be calculated from the time difference between the receipt of the two signals.

The GPS satellite transmits on two frequencies, and the use of the two results in errors of about ten meters. However, vertical accuracy of ten meters on an approach to landing is far from sufficient. A system capable of enhancing accuracy to fractions of a meter is called *differential GPS.* A number of differential navigation systems have been employed over a forty-year period, including differential LORAN and differential OMEGA. Both of these systems stood to benefit from differential enhancement as they were not stellar performers in the area of accuracy.

A differential navigation system uses a receiver at a known place called a *reference receiver*, or, in the case of GPS, the *reference locator*. Essentially the role of the reference locator is to receive the actual signals and determine how those signals differ from what is expected. Because the reference locator "knows" where it is located, the receiver knows what the correct signals should be. The differences between the actual received signals and what should be received are used to calculate error corrections. These corrections are transmitted via a data link to users, who then correct the errors in calculations. Using a differential system with GPS, an accuracy of better than one meter in all three axes is possible at all times and in all kinds of weather. Using GPS for landing and en-route navigation means that a "seamless" navigation system can provide continuous worldwide navigation from takeoff to landing.

As previously mentioned, one of the more important aspects of GPS navigation for all phases of flight is that every airport can be an instrument field. With the addition of a differential reference locator, the GPS can be used for a CAT III approach. Instrument approaches do not have to be made to an airport. Instrument approaches can essentially be made to anywhere. This may not appear to have any application as most aircraft cannot land just anywhere. That is not true for all aircraft, however. In the case of rotorcraft, precision instrument approaches may be made to rooftop helipads at hospitals and so on.

At the time of this writing, two differential GPS systems are being designed and will be installed in the United States. These two systems are called the *wide-area augmentation system*, or WAAS, and the *local-area augmentation system*, or LAAS. The WAAS will cover a large geographical range and permit moderate precision landings at a wide range of airports. The LAAS provides more precise correction and is intended to be used for one large airport or a close group of several airports.

One of the most important characteristics of the GPS system is the low cost of the user equipment. No navigation system has ever been introduced with so much performance for such a low price. This is particularly important for general aviation aircraft. It is possible for GA aircraft to have navigation performance on a par with the largest aircraft. The modern GPS receiver is small and light and consumes very little power. It is common nowadays to find a battery-operated, portable GPS receiver aboard a glider or free balloon.

Automatic Flight Control

As good as laser and fiber-optic gyros are, they must be corrected occasionally due to drift. The drift rate of a modern inertial navigation system is very small—only a fraction of a nautical mile per hour. But it is never zero, and the error increases with flight time.

The GPS system, on the other hand, never drifts and is always accurate. Unlike the inertial navigation system, the GPS takes a few seconds to determine a fix. The inertial navigation system, however, can respond to a change in position nearly instantly. Furthermore, the inertial navigation system can provide heading and attitude information, whereas the GPS alone cannot. If the inertial navigation system and the GPS were combined, we would have the best of all worlds. We would have the unchanging accuracy of the GPS coupled with the spontaneous response of the inertial navigation system. If we were to use the inertial navigation system and the GPS to automatically control the aircraft, we could smooth out the bumps of rough air using the quick response of the inertial navigation system while maintaining precise navigation even for long trips.

Inertial navigation and GPS are the core of the modern, flight-management system, or FMS. This system can automatically control an aircraft in all "six degrees of freedom," or latitude, longitude, altitude, roll, pitch, and yaw. The FMS is programmed with waypoints, which include latitude, longitude, and altitude. The FMS also receives inputs from the VOR and ILS systems so that victor airways can be flown from VOR information and instrument approaches may be flown from ILS information. The FMS can be used to hold a steady climb or descent or to hold a heading. "Hand flying" an airplane is nearly a thing of the past. Just as soon as an aircraft has taken off, the FMS flies the aircraft.

Further Reading

Abzug, Malcolm J., and E. Eugene Larrabee. *Airplane Stability and Control: A History of the Technologies That Made Aviation Possible.* Cambridge, New York: Cambridge University Press, 1997.

Augustine, Norman R. *America's Space Program: Wishing upon a Star or Going to the Stars?* Washington, D.C.: Smithsonian Institution Press, 1992.

Augustine, Norman R. *Augustine's Laws.* New York: Penguin Books, 1987.

Bilstein, Roger E. *The American Aerospace Industry: From Workshop to Global Enterprise.* New York: Twayne Publishers, 1996.

McCurdy, Howard E. *Space and the American Imagination.* Washington, D.C.: Smithsonian Institution Press, 1997.

Thornborough, Anthony. *Sky Spies: Three Decades of Airborne Reconnaissance.* London: Arms and Armour Press, 1993.

Titterton, David H. *Strapdown Inertial Navigation Technology.* London: Peter Peregrinis, Ltd., 1997.

Chapter 7

The Computer Age
A Small Chip Starts a Big Revolution

In 1971 a new integrated circuit was introduced to the electronics industry that would change the world forever. It is hard to imagine that a single integrated circuit would have an effect on the way we all live, but it did. The chip was the Intel 4004 microprocessor, which marked the first step toward putting a computer on a chip. The 4004 was extremely crude by modern standards and required a number of peripheral chips to do any computing with the device. Therefore, the 4004 was not a microcomputer but only a part of the computer: the processor. The chip contained about two thousand transistors, which was quite a feat for the time since, only a few years before, the state of the art for the number of integrated circuits in a chip was perhaps ten transistors.

The 4004 was a four-bit device, which means the chip operates with four-bit data words. As an example, numbers from zero to fifteen could be represented with four binary bits. One of the biggest problems with the 4004 was that it was in a sixteen-pin package, which means that there were only sixteen electrical connections to the chip. A microprocessor must handle input data, output data, and addresses. With only sixteen pins it was not possible to assign unique pins to these three binary words. The data and addressing were fed to and from the chip by a technique called *multiplexing*. The pins of the 4004 were shared between data and addresses by placing addresses on the pins at one time, input data at another time, and output data at yet another time slot. Other signals that must be extracted from the microprocessor involved timing and the status of the chip, which seriously limited the speed of the chip.

How can a four-bit microprocessor perform any type of calculation if the chip can handle numbers from only zero to fifteen? When we add two decimal numbers together, we do it one digit pair at a time. If the sum of two digits exceeds nine, the largest decimal digit, we generate a "carry," a number that is "carried" over to the next digit addition. Thus, the result of a decimal addition is a digit from zero to nine and a possible carry. The carry is only a zero or one, and therefore it is a carry "bit." When the next two digits are added, the carry is added to them. This is exactly what children are taught in school. Subtraction is the same as adding. There are a number of techniques that use addition to implement subtraction as well as multiplication. Division is the most difficult mathematical operation to implement with a computer. Most of the early programs were carefully written to avoid division and, most important, to ensure that a calculation never attempts to divide by zero. Dividing by zero caused some early computers to "hang up," a term we will discuss later.

In spite of the first microcomputer's limited capabilities, the device could replace dozens of discrete, integrated circuits. The microprocessor was a programmable chip. That means that the basic microprocessor would do what you programmed it to do. As an example, if the microprocessor is to calculate ground speed from distance information, the processor is programmed to perform that function. Or perhaps the "time to go" is desired from the ground speed and distance. The same processor may be used with a different program.

The 4004's clock was 750 kHz and was restricted to adding only one decimal digit at a time. By modern standards this processor was extremely slow. Modern microprocessor-based computers, by comparison, have clocks that run in excess of one thousand MHz, or one gigahertz (GHz), at the time of this writing. This is more than a thousand times faster than the 4004. Also, modern computers handle sixteen and thirty-two bits at a time and do not require multiplexing. When modern data-handling techniques are employed, the newest microprocessors are on the order of a million times faster than the 4004.

Not all applications of the microprocessor are mathematical. One of the more important applications was controlling things. Specialized microprocessors called *microcontrollers* became available early in the history of the microprocessor. A good example of a microcontroller is in the keyboard of a personal computer. The keyboard has a dedicated computer inside that first determines which keys are depressed and then transmits encoded information to the computer. One of the first applications of the microprocessor was controlling the front panels of avionics systems. As early as 1977 the microprocessor was controlling the functions of communications and navigation equipment in small aircraft. It was now possible to store often-used frequencies and navigation data in a memory and retrieve the desired data when needed. By modern standards this is a trivial application of a computer, but in 1977 this was a very valuable capability for aircraft and did much to reduce pilot workload. As the first to use the new microprocessor, general aviation was way ahead of military and air-transport aircraft. This was because of some of the nasty habits of the new microprocessor.

Hang-Ups

Today everyone is familiar with computers and some of their unpleasant habits. Anyone who has used a personal computer (PC) for any significant time has experienced what is called a "hang-up." This is what happens when a program does not respond to the normal inputs, such as the keyboard or mouse, and simply stops working. The computer must be reset and the program restarted. This is so common that operating systems have restart commands to minimize the disruption of a hung-up program. As a final resort there is a "reset" push switch on most computers, which is the "hardware" or "hard" reset.

What causes these types of failures? There are a number of reasons, and most are very difficult to detect or predict. First is a software problem. To simplify a complex subject as much as possible, if a program takes a particular path, the program will become derailed, so to speak, and will wander off and not return. A computer is like following a map. You go from one place to another performing various tasks, and somehow there must be a path back to "home base," where you receive new assignments. Every task has a road sign that says "this way home." Imagine that if you took a certain path and you did not find a road sign or you found improper road signs that directed you in a never-ending circle.

In the early days of microprocessor-controlled equipment, if an operator held a push button while turning another switch only while in one particular mode of operation, the system would "crash," or hang up. This unusual path led the computer to an endless circle and trapped it. Or, if the computer were directed to an area where there were no "this way home" signs, it would essentially wander around, lost in cyberspace. Problems such as these were very difficult to find because the operators could never remember exactly what they were doing when the unit crashed. Other hang-ups were caused by short-lived electrical signals called *glitches*. These would cause a miscommunication within the microprocessor, so the processor would incorrectly execute a command, and the program would crash.

When a program crashed, the microprocessor would usually execute what was called a *short loop*. By "loop" we mean the microprocessor was executing the same commands over and over. If one were lucky, the loop did not create much havoc. However, if the loop was long and the commands executed involved visible results (such as lighting lamps and activating devices), the computer appeared to have gone berserk. These types of failures were a new experience for those in the electronics industry. The discrete logic that had been used before generally failed in a rather innocuous way by just quitting all together or becoming unresponsive to inputs. It was rather disconcerting to see a front panel of a system with all of the lamps blinking and the displayed digits dancing across the panel in undecipherable gibberish. One thing was clear: The appearance of the words "glitch," "crash," and "berserk" as used in the description of avionics failures was not welcomed in the aviation community.

The air-transport-industry avionics are well-proven technologies that meet strict requirements. General-aviation avionics tend to be more state of the art. General-aviation buyers are more likely to purchase a new technology on the assumption that if an avionics system fails during flight, there are alternatives to resolve the problem. As an example, most general-aviation aircraft fly visual flight rules. If a general-aviation aircraft loses all navigation radios during flight, the pilot assumes visual navigation and either returns home, completes the trip, or lands at the nearest airport and has the radios fixed. If the trip is aborted, that is an annoyance, but it is preferable to getting lost and creating a serious problem.

The same scenario for an air-transport aircraft may bring about quite different results. First, large aircraft fly instrument flight rules, IFR, regardless of the weather. Losing avionics would preclude that aircraft from legally flying IFR even though the weather may be perfectly clear. If the trip is not completed, the airline stands to lose considerable revenue, even if there is no danger to the passengers.

A number of techniques were developed to prevent program hang-ups in microprocessors. First was improved software. It was not possible to manually operate every keyboard switch, turn every knob, and simulate every possible operational mode of an avionics system to test for hang-ups. It was possible, however, to use test software to test the system software. The test software would produce large numbers of possible inputs and test the software for glitches. Hardware glitches were much more difficult to test, but improvements were made in this area as well. One technique to eliminate hang-ups was the "watchdog" circuit. This was a hardware circuit that monitored various signals within the computer. When activity ceased for a period of time (such as ten seconds), it was apparent that the computer was hung up, so the watchdog would hardware reset the computer. This process usually caused some temporary effects and did not go unnoticed, but it was certainly better than becoming hung up.

The Intel 4004 was followed by the 8008, which was an improved eight-bit processor containing about four thousand transistors. It had a lot of the same disadvantages of the 4004 since it was in a sixteen-pin package. However, it was more than twice as fast as the 4004 because it could handle twice as many bits. Later, in 1974, the 8080, the first in a long line of Intel microprocessors, appeared. This chip contained more than eight thousand transistors, more than any single device before it. Amateur computer users appeared using several of the available computer kits such as the Altair and the Imsai. Well-known names began to appear: Jobs, Wozniak, Gates, and others. The microprocessor era was accelerating at a fantastic pace.

The growth of microcomputers and in particular microprocessors was fascinating to watch. Gordon Moore, one of the founders of Intel, predicted that the number of transistors contained in a microprocessor chip would double every eighteen months. This became known as Moore's law; the subsequent development of microprocessor technology proved Moore correct. At the time of this writing the largest microprocessors contain nearly ten million transistors. It is possible that Moore's law has run its course as the largest integrated circuits made today are not microprocessors but *programmable gate arrays*. These chips are programmable hardware. In fact, these devices can be programmed to be a microprocessor. There is no doubt that these programmable gate arrays will play a very important role in the future of avionics.

Testing Software

Early microcomputers were programmed in what is called *assembly language*, which was a very tedious process. Assembly language was as close to directly talking in machine language as one could get. However, after the mnemonics (that is, the most basic instructions) were typed in, the program had to be converted to machine language with a process called "assembling." It could take thirty minutes or more to assemble a program that would do nothing more than operate the front panel of a sophisticated avionics system. If assembly was unsuccessful, error messages would appear, such as "three errors," but the message did not specify where or what type of error had occurred; programmers were on their own. In spite of all of this difficulty, microprocessors were being applied to all types of airborne electronics.

More sophisticated, higher-level languages appeared and made programming easier and more efficient. However, the higher-level languages moved the designer farther away from the actual operation of the microprocessor and made hang-ups more difficult to find and fix. Fortunately, new software tools were also becoming available that would find the errors and direct efforts at fixing them.

The microprocessor introduced a new dimension to avionics equipment test and evaluation. The RTCA and the FAA had no methods of evaluating software for potential problems. When an avionics system was tested for FAA certification, the equipment was operated and tested for abnormal voltages, currents, temperatures, altitudes, vibrations, and so on. If the equipment continued to function and meet the required performance levels, it passed. When avionics began using microprocessors, the tests were the same. The tests were "box-level" tests. That is, it did not matter what was inside of the box as long as the equipment met the required performance specification. However, engineers knew that pressing a sequence of switches or entering inputs from other equipment could crash the software and bring down some equipment that would pass the conventional tests. A new test had to be formulated that would test software.

The FAA requested that a special committee of the RTCA be empanelled to draft a document for software testing. Special Committee 145 produced document DO-178, "Software Considerations in Airborne Systems and Equipment Certification," which was released in November, 1981. Testing of software involved more than determining the software's resistance to crashing. The actual performance of the equipment was set by the software. Most RTCA performance standards were written before the microprocessor, so the operation of the tested equipment was very simple. It was possible to make a good assessment of the performance of an avionics system by turning a few knobs and reading a few meters. The new microprocessor equipment had many modes of operation and could provide many more data. It was not possible to exercise all of the capabilities of a sophisticated, microprocessor-controlled system manually. Therefore, DO-178 addresses the testing of software as software rather than hardware. The only way to completely test a software function is with a software test routine.

Improved and much more powerful microprocessors quickly followed the 8080. The 8085 was a true, single-chip processor and was in a forty-pin package that did not require as much multiplexing as the 4004 and 8008 did. The 8085 integrated several of the peripheral chips from the 8080 and was much easier to use.

Intel was not the only supplier of microprocessors. Motorola produced its first in a series of 6800 processors. Advanced Micro Devices, AMD, made Intel "look-alikes." All the major chip manufacturers—Texas Instruments, National Semiconductor, and a host of foreign chip makers—had their own brands of microprocessors.

Shortly after the 8085, the 8086, the first in a well-known series of chips, appeared. The 8086 was the processor used in the first IBM PC and was the first of the series that included the 8086, 80186, 80286, 80386, and 80486, which are known to most computer users. The introduction of the IBM PC had an incredible effect on all industries. Before IBM, only the technical hobbyists, the computer amateurs, had personal computers. Industry used terminals that were connected to a mainframe computer. With the introduction of the IBM PC, huge amounts of software became available, including software development for microprocessors. The microprocessor revolution was in full swing.

By the mid-1980s the front-panel functions of almost all avionics were controlled by microprocessors. Simple navigation systems now boasted all types of features. Waypoints could be stored before a trip and recalled without the need to look up and enter the data into the navigation system. The sophisticated front panels of the new avionics with microprocessors reduced the crew's workload. This was only the beginning. Later systems would make it possible to eliminate the flight engineer's job in large aircraft and also make possible the single-pilot, business jet aircraft. This is a very important fact. Because of electronics, one-third or one-half of an aircraft's crew could be eliminated. Because of electronics, a number of crew functions disappeared from aircraft, such as the tasks previously performed by radio operators, navigators, flight engineers, and bombardiers.

Networks on Board

When most people think of computers and the tremendous capability they offer, they think of the Internet. This is interesting in that it is not computer technology that makes the Internet possible but rather communications technology. A similar situation developed in the evolution of the microprocessor in aircraft. One of the first applications of the microprocessor, as previously mentioned, was the controlling of the front panels of avionics systems. This was extended to controlling several units from one panel. If the typical aircraft had two navigation receivers, which operated two distance-measuring sets, and two communications receivers, why not control all four receivers and the distance-measuring equipment from one panel?

To do this, the front panel had to communicate with the receivers. The navigation receiver or communications transceiver typically has its own microprocessor, which communicates with another microprocessor or, more precisely, a microcontroller in the control panel. Early in the development of microprocessors, it was recognized that communications between processors would greatly enhance the applications of the devices. Even the early microprocessors had the built-in capability of communicating with other microprocessors and other circuits, which is called *networking*. Aircraft began to have several networked computers. A modern air-transport aircraft can have hundreds of networked computers.

There were even more fantastic applications. Why not take all of the navigation information from all of the equipment on the aircraft and use this information to navigate? This technique is called *random navigation,* or RNAV (the "random" meant the aircraft could fly anywhere, as opposed to only on an airway, using information from VOR, DME, LORAN-C, Omega, and GPS). There was communication from every navigation system aboard the aircraft. The aircraft navigation was a single system rather than redundant, stand-alone navigation sets. The performance of the system approach was fabulous and extremely versatile. On the other hand, if the controlling unit failed, the whole system would fail. This indicated the need for a system characteristic called *fault tolerance.* This implies that one function can fail, but that failure will not cause the entire system to fail.

Computers in Flight Control

There was always reluctance to use the microprocessor where it would eventually find its most important application: flight-control systems. Usually new technologies appear in general aviation first, as previously discussed. After it can be demonstrated that the new technology is trustworthy, it is moved to the air-transport arena.

There was extreme reluctance to use the microprocessor in an autopilot, even in general-aviation aircraft. Finally, in the early 1980s the microprocessor found its way into flight control. A flight-control system is a very difficult design task. Not only does the system accept huge amounts of data from a variety of sensors, it must also direct a number of electric motors to control the aircraft. One of the more difficult aspects of a flight-control system is that the system must be customized for each aircraft type.. Unlike a navigation receiver or communications set that can be installed in any aircraft, the flight-control system must be custom adjusted to the characteristics of a particular aircraft.

In the older, analog-control systems, this customization was performed by changing the values of components in the equipment. This meant that a specific part number was required for each aircraft. The units had to pass through a different production line and be stocked. The importance of the microprocessor is that the chip can be programmed, and thus only one flight-control system can be made and stocked. When a system is ordered, the unit is programmed for that particular aircraft before shipping. The irony is that after initially being rejected for flight control, the microprocessor was the driving force behind huge changes in this aspect of flight. The purpose of a flight-control system is to control the aircraft's position in all six degrees of freedom.

The type of navigation in use determines the path of an aircraft. In the case of simple navigation, such as holding a particular heading and altitude, the flight-control system receives input information from the directional gyro and the altimeter. If a radio navigation aid is used, then the controlling information will come from the radio navigation system. When an aircraft is being controlled to maintain a specific heading, the autopilot is in the "heading-hold" mode. Likewise, when altitude is being controlled, this is called the "altitude-hold" mode. When the pilot wishes to change heading or altitude, the control panel for the autopilot has a turn knob and a climb/descend knob. Rotating the turn knob causes the aircraft to turn at a fixed rate. The climb/descend knob causes the aircraft to climb or descend at a constant rate. When the aircraft has reached the desired heading or altitude, the turn or climb/descend knob is reset to zero, and the aircraft holds the resulting heading and altitude. When the aircraft is in the heading- or altitude-hold mode, it can be flown completely with the turn and climb knobs. If the autopilot can be turned on at all times, why couldn't the yoke and rudder pedals be replaced with electronic knobs? The answer is "they could," which is the concept behind "fly by wire."

Although faith in modern microcomputers is growing, there is always the very small possibility that a computer, other hardware, or software can fail, causing a calamity. Because of this, various extreme safeguards are taken when an aircraft is controlled by a fly-by-wire system. Hardware failures are easy to understand. A part burns out and fails. What is important is that the failure usually remains visible until the repair technician can find and fix it. As discussed previously, software failures are more difficult to understand. Often a software failure will not remain to be found. Once the equipment is turned off, the failure disappears and cannot be recreated. However, as airborne computer systems involve more complex software, the likelihood of a software failure may become greater than hardware failure.

When a computer affects the safety of an aircraft, safeguards against computer failure are required. One such safeguard is *redundancy*. Redundancy is the use of more than one computer to control the aircraft. If one computer fails, the remaining computer takes over the task, and no harm is done. However, how does one determine which of the redundant computers has failed? A technique employed to solve this dilemma is to use three computers operating simultaneously. The computers are fed the same input information, and the output results are compared. If one computer fails, its output results do not agree with the results from the remaining two. What if the computers are identical and a software problem causes an error at the output? If the input data presented to the computers are the same, and they should be, all three computers make the same mistake and all three agree. This results in an undetectable error.

What would happen if two of the computers were identical? Now the problem would appear in two of them, so the situation now is that two computers agree and the third is different and assumed to be at fault. In this situation the two errant computers are assumed to be correct, while the computer that is operating correctly is assumed to be errant. When redundant computers are used, none of the computers can be the same. This means that they cannot share any significant component, hardware, or software. They should use different microprocessors. The software should be written using different languages. Different design teams should be used, and those teams should not share information other than the design specifications. This is because if one group shares with another group a simple algorithm, for instance, that just happens to fail under certain conditions, this bug will be infused into the other computers. When conditions are right and the algorithm is used in the program and fails, all of the computers fail in the same way and thus the failure escapes detection.

The most important tool for ensuring failure-free computers is time and experience. Computer algorithms and code that have been shown to operate for years without problems are used in future designs. Once a computer has been operating in the field and the program bugs have been found and fixed, the computer's reliability is as good as that of any other electronic system.

Controlling Engines

Electronic systems have been used since the 1970s for certain engine controls. One of the first applications was to synchronize the engines of multiple-engine aircraft to eliminate the annoying "beat" effect from two engines turning at different rates. This type of system was a throttle-control system in which the throttle of one of the two engines (in the case of a twin-engine aircraft) could be adjusted slightly with an automatic control system. The key word here is *slightly*. The pilot was required to manually set the throttles to nearly synchronized, and then the automatic system would maintain perfect synchronization.

Because of the historic distrust of electronic control systems, there is something reassuring about a mechanical linkage between the throttle lever in the flight deck and the engine on the wing. In large aircraft, a direct mechanical linkage is not possible, so a rudimentary "electric" control system is used. Once the fear of the microprocessor subsided, a digital, engine-control system was acceptable. Such a system is called a "full authority, digital-engine controller," or FADEC. "Full authority" means that, unlike the early prop synchronizer that could adjust the throttle a small amount, the FADEC has the ability to control the engine from idle to full power.

The FADEC is more than just a throttle. In reciprocating-engine, propeller aircraft, for example, the FADEC would replace the throttle, prop pitch, carburetor heat, mixture, and primer with one control simply called "power." In the final analysis, it was not uncommon for a pilot to incorrectly set these controls and damage an engine in flight or not have the power required when needed. Here is another way of looking at the situation: The last automobile to have a "mixture control" was the model T, but aircraft engines incorporated this device for another seventy-five years.

In this section we have "painted" the "picture" of modern aircraft, which are flown by electronics. The engine control in the cockpit sends signals only to an engine-control computer, which communicates with digital signals to the engines. The navigation is from a computer, which takes data from all the navigation equipment aboard the aircraft and provides a unified "navigation solution." And that's not all. In later sections we will investigate computer displays in the cockpit and data links to the world.

Crowding of the Skies: Crowding of the Airwaves

By the mid-1980s the steady growth of air traffic, both commercial and general aviation, increased the load on the air-traffic, control-radar beacon system, ATCRBS. This system is the secondary radar system developed from the World War II IFF transponder. It provides aircraft identity and altitude to the air-traffic controller. In the heavily traveled terminal areas, particularly Los Angeles and the east-coast corridor, false targets and missing targets were becoming too commonplace. An improved secondary radar system was required. With the exception of an increase in the number of codes handled by transponders (from 64 to 4,096 in 1962), the ATCRBS was essentially a World War II system. There were many improvements made on both the airborne transponders and the ground interrogators, but the basic concept was forty years old in the mid-1980s. One of the major problems was that there was insufficient capacity for every aircraft in the world to have its own unique code. Of the possible 4,096 codes, some are reserved for special functions such as flight using visual flight rules (VFR) and emergencies. Therefore, the codes are "recycled." In other words, once an aircraft has left an area, the code that aircraft was using can be reassigned to another aircraft.

Second, all aircraft that received an interrogation would respond to it. If two aircraft were nearly the same distance from a ground interrogator, the reply from both aircraft would arrive at the ground interrogator at the same time and mutually interfere, causing "garble." One solution to this problem would be to individually interrogate transponders with their own unique address, and thus only those interrogated would reply. In order for this to be implemented, every transponder would have to have its own unique address, which was not possible with the ATCRBS system.

A new transponder system was developed called the "mode S" transponder, where "S" stands for "select"—the ability to selectively interrogate a transponder. The name was an extension of the ATCRBS transponder's two modes of operation: mode A for identity, mode C for altitude, and now mode S for select. This transponder was developed by a number of companies including Bendix (later Allied Signal, now Honeywell), Rockwell-Collins, MITRE Corporation, and MIT Lincoln Labs. The requirements were hammered out in RTCA SC-142 with the final document, DO-181, released in May, 1983.

The mode-S transponder brought air-traffic control's secondary radar to the latest technology. In fact, one of its major problems was that it was a bit too complicated. It was such a huge leap in technology that avionics engineers and technicians had a hard time grasping the system's operation. Fortunately, with the explosion of the Internet and sophisticated digital communications systems, after a few years the mode-S transponder did not appear so awesome.

Unlike the old ATCRBS system, the new mode-S system was capable of data transmission, control of transponders from the ground, and message broadcasting. Some of the latest digital techniques were used, such as a sophisticated "cyclic redundancy check" (CRC), or error detection. Complex transmitters were proposed for the airborne units. The ground stations used even more complex encoding and transmission techniques. Expensive ground-station transmitters are not a problem. A radar installation is by definition expensive, but the government purchases them. Expense was a problem when the airborne units were considered, however. If aircraft owners balked at the high cost of transponders, it would be difficult to ensure widespread installation of the new mode-S unit. If a large percentage of the aircraft fleet was still using the older ATCRBS transponder, garble would remain a problem.

During the RTCA meetings, these problems between the general-aviation manufacturers and the scientists at the "think tanks" were smoothed out, and a workable system was defined. As it turned out, the FAA mandated the installation of mode-S in all air-transport aircraft, both new and in service. For general aviation, a schedule over a period of about seven years would at first require mode-S transponders in new aircraft, then in aircraft being fitted with new avionics, and finally would place a total ban on the manufacture of the older ATCRBS transponders. The plan was so unpopular with general-aviation aircraft owners that the FAA partially rescinded it.

Electronic Collision-Avoidance Systems

In the postwar period, this was almost guaranteed: If a midair collision occurred, there would be suggestions and discussions relative to an electronic collision-avoidance system. Midair collisions are very rare, but when they occur, they are the most costly of all aircraft accidents. There have been a number of midair collisions that sparked discussion of a collision-avoidance system. The first was the June, 1956, collision of a TWA Lockheed Lo49 Super Constellation and a United Airlines Douglas DC-7 over the Grand Canyon, with a loss of 128 lives. At that time this collision caused the greatest number of fatalities of any aircraft accident. In April, 1958, a U.S. Air Force F-100 collided with a United Airlines DC-7 over Arden, New York. Two years later, two aircraft, both under instrument flight rules and both making an approach to Idlewild Airport, now JFK, collided over New York City in December, 1960. The aircraft involved were a propeller-driven Super Constellation and a jet-powered DC-8. In July, 1967, a Piedmont 727 collided with a Cessna 310 over Hendersonville, North Carolina.

The midair collisions outlined here are not the only ones that took place over that eleven-year period, but they demonstrate some important characteristics. First, not all of these collisions occurred in the terminal area. Several collisions were between small aircraft and larger air-transport planes. Military aircraft were involved as well. Some of the aircraft were under air-traffic control. It is essential that any collision-avoidance system protect small aircraft, large aircraft, and military aircraft and be effective both in the terminal area and en route.

Before an effective collision-avoidance system can be designed, one of the first questions is, who is responsible for the safe separation of aircraft? When under air-traffic control, aircraft fly only where the controller directs them, but collisions of aircraft have occurred while the planes involved were under positive control. Then there is the problem posed by an aircraft that is not under air-traffic control. All possible precautions are taken to prevent aircraft that are not under air-traffic control from mixing with controlled aircraft. Air-traffic control is a collision-avoidance system already, but despite this system, midair collisions have occurred. Air-traffic control is not a perfect system, and there is the possibility that controlled and uncontrolled aircraft as well as two uncontrolled aircraft may vie for the same airspace. Therefore, electronic, collision-avoidance systems are airborne, with each aircraft responsible for its own protection.

One of the first attempts at an airborne, electronic-collision avoidance system was in 1956 with a system called airborne collision avoidance system, or ACAS (pronounced a-kass). This system used microwave transponders different from the ones used for air-traffic control's secondary radar. The system determined range by interrogating transponders of other aircraft. Then it calculated the range by measuring the time between the transmission of the interrogation and the receipt of the reply. The system was an airborne secondary radar with no associated primary radar. The major disadvantage of this and many other collision-avoidance systems is that all aircraft must be equipped with collision-avoidance equipment. This requires government-mandated equipment and restrictions on foreign aircraft operating in U.S. airspace.

The key to ACAS was a parameter called τ, the Greek letter tau. This parameter was used by Dr. John "Smiley" Morell from the Bendix Corporation. Dr. Morell's 1956 publication, *Fundamental Physics of the Aircraft Collision Problem,* was one of the first scientific investigations into midair collisions and their prevention. Tau was the measured range to a target aircraft divided by the range rate. *Range rate* is the rate of change of distance per second. As an example, assume the range to an aircraft is determined to be 2.15 nautical miles, and then the next time it is measured at 2.10 nautical miles. If the time between the measurements is 5 seconds, the range rate is 0.05 nautical miles divided by 5 seconds, or 0.01 nautical miles per second. Tau for this example is the range, 2.10 nautical miles, divided by the range rate of 0.01 nautical miles per second, which is 210 seconds. Notice that when range is divided by range rate, the result is time. Tau is the time between when a distance measurement is made and when a collision will occur if the situation remains the same for that time.

A few things must be understood concerning tau. First, if the range is increasing, the distance between the two aircraft is increasing, so there is no collision problem. Second, in many situations tau will indicate a chance of a collision when there is actually none.

Consider a situation in which two aircraft at different altitudes pass directly overhead with a separation of, say, 1,000 feet. Such a scenario is not a collision situation. But, if the 1,000-foot separation is horizontal rather than vertical, then the situation is too close for comfort because aircraft have very low velocities in the vertical direction. Assume an aircraft is climbing at 1,000 feet per minute, which is quite common. This is a velocity of slightly less than 10 knots. Compare this to the case of two aircraft approaching head on at a typical airspeed of 400 knots or a closing velocity of twice this, or 800 knots. These two aircraft are approaching at a rate of 1,350 feet per second.

Another scenario might involve aircraft at the same altitude that have crossing paths that will take one aircraft in front of the other at a distance of one nautical mile. The problem of using tau as an indicator of potential collision is that tau is a scalar, that is, only a number with no reference to direction. Tau is time, and time is never a vector. Tau was also calculated from two parameters that were not vectors: range and range rate. However, both range and range rate can be made to be vectors if more information is available.

Collins Radio had submitted a proposal to design and produce the collision-avoidance system. Collins continued to provide avionics after the war, had become one of the major suppliers of air-transport avionics, and enjoyed a stellar reputation. The potential of the contract was for more than ten million dollars—no small sum in 1956. The company estimated the job would take between two and four years. Even though Collins Radio was known for designing sophisticated systems in record time, the company struggled with the system. In spite of the tremendous advances made in electronics during and after the war, the challenges were more than the state of the art could solve. Arthur Collins, the company's founder, appreciated the need for airborne-collision avoidance but also eventually understood that the 1956 technology was incapable of producing a workable system. In a move that few companies have ever contemplated, Collins returned all of the up-front money it had spent on the project and told his customers, most major airlines, to wait before undertaking a collision-avoidance system.

The ACAS system failed for two reasons. First, it was very expensive, and second, ACAS required installation of ACAS equipment on all aircraft to be effective. The FAA has mandated that certain equipment be installed in aircraft on a number of occasions. Such equipment includes emergency locator transmitters, transponders, cockpit data recorders, and flight data recorders. Every time the FAA has mandated a system, their directive has been met with great resistance. One technique the FAA used is to limit the airspace in which an aircraft can operate without a particular piece of equipment. The transponder was not mandated for all aircraft, but the airspace in which an aircraft not equipped with a transponder could operate was limited. In many parts of the United States, the airspace that a nontransponder-equipped aircraft could operate in became so restricted that the aircraft was virtually unusable. Mandating a very expensive collision-avoidance system for all aircraft would not work.

Another early system was called ERoS, where the "o" is not the letter *O* but zero. ERoS was Eliminate Range Zeros ("range zero" meaning "collision"). This system failed for similar reasons as ACAS.

A later collision-avoidance system was BCAS (pronounced bee-kass), where the *B* stood for "beacon," more precisely, the beacon of the air-traffic control, radar beacon system, ATCRBS. The advantage of using the ATCRBS transponders was that many aircraft were already equipped with them, and the FAA had set a precedent in mandating the use of transponders in certain airspace.

The BCAS system had a number of variants. Some received signals only from nearby transponders. This made measuring range difficult. In a very simplistic way, strong signals were received from the closest transponders. Radio signals decrease in power inversely with the distance squared. Since transponders were limited in both their minimum and maximum power, it was possible to get a rough estimate of the range to the transponder. The problem was that the estimate was a bit too rough. The alternative was to interrogate nearby transponders and measure an accurate time delay, just as a ground radar station. This variant of the BCAS system was called the "active BCAS."

A number of concerns arose with regard to this type of system. First, every aircraft would be an interrogator. This would increase the amount of interference and could compromise the ATCRBS system. To counter this it was necessary to restrict the airborne interrogators to their maximum range and how often they can interrogate. If the range of the airborne interrogators is limited to less than twenty nautical miles and the rate of interrogation kept to only a few every second, it may be possible to add the airborne interrogators of the BCAS system without seriously jeopardizing the secondary radar system. An important change in the air-traffic-control radar system was the introduction of the mode-S transponder and interrogators. The mode-S transponder had a number of capabilities that would improve the situation when a collision-avoidance system came on line. In addition, the mode-S transponder could become a part of a collision-avoidance system.

An RTCA special committee, SC-147, was formed to investigate the BCAS collision-avoidance system. As it is with voluntary committee activities, progress was slow. Then in 1978 a midair collision occurred over San Diego airport between a Pacific Southwest Airways Boeing 727 and a Cessna 172, with a loss of 135 lives. Shortly afterward, FAA administrator J. Lynn Helms sent a letter to SC-147, wherein he said he wanted the definition of a collision-avoidance system within a year. Helms offered the FAA's cooperation, funding, and legislation to mandate an airborne-collision avoidance system. To make it clear he was serious, the collision system was renamed from BCAS to TCAS (tee-kass), for traffic-alert and collision-avoidance system. After three decades of fits and starts on a variety of collision-avoidance systems, a system was finally going to materialize.

The task was not going to be easy, however. Designing, specifying, making, or any activity associated with a collision-avoidance system had a scary element to it. What manufacturers feared was litigation, should a collision-avoidance system fail to prevent a collision or—worse—cause a collision in what should have been a safe situation. Before TCAS, the safe separation of aircraft under air-traffic control was the FAA's responsibility. Should an unfortunate incident take place, the responsible party would be the government, which was protected from bankrupting litigation. Also about this time, the pitfalls of software were being researched, and the RTCA issued its software-certification document, DO-178.

The airborne TCAS equipment consists of a secondary radar interrogator, a method of display, and a computer. There was no question that TCAS would require a significant amount of computing. Of all the possible failure modes, the greatest concern was for software problems that might cause dangerous situations.

No manufacturer wanted to be responsible for writing the collision-avoidance software. A new policy was introduced to the TCAS committee and has never been repeated. The FAA would hire a private company to write the software for the TCAS, thus making the government the owner of the software. All of the algorithms for the TCAS computer were provided to the manufacturers in the form of "pseudocode." This was a detailed skeleton of the code that could be modified to fit whatever processor the manufacturer chose.

As it turned out, the pseudocode had problems. Flaws in the software appeared. Fortunately, none of them caused any accidents, and the software problems were eventually identified. Normally, as soon as a manufacturer discovers a serious flaw in software, the engineering department determines the fix. Then notices are immediately sent to owners of the equipment, requesting that the customer have the TCAS serviced at a dealer. In the case of pseudocode, the manufacturer finds a fix for that company's software, but a generic fix must be configured for the pseudocode, and that fix must be approved by the special committee and then sent to the manufacturers. One of the tenets of pseudocode is that all of the manufacturers of TCAS must use the same basic style of software. The disadvantage of this is the overwhelming inertia behind each change—that of the entire industry. Since the introduction of DO-185, the document outlining the TCAS system (so lengthy it consists of two volumes), there have been seven changes at the time of this writing, mostly for software problems.

The TCAS system interrogates secondary radar transponders within about twenty nautical miles with an altitude interrogation. The altitude information eliminates about ninety percent of the replying aircraft as potential threats. By measuring the time delay between the interrogation and the reply, the range of the aircraft is determined. The range measurement eliminates a considerable number of the aircraft at the same altitude as they are simply too far away to be an immediate threat. The process of eliminating from computation those aircraft that are not an immediate threat is simple. Within the rather short range of twenty nautical miles, however, there could be a large number of aircraft, particularly in the terminal area. The aircraft that remain to be tracked are those that are relatively close to "own aircraft" and at own aircraft's altitude or climbing or descending to that altitude. These aircraft are called "proximate" and are monitored by the system.

Tau is calculated on each proximate aircraft. Once tau decreases to less than about forty seconds, the aircraft is called an "intruder," and a traffic advisory is activated. This is a voice announcement of the word "traffic" three times in the crew's headphones and the flight deck's speaker.

In its simplest form, the TCAS system gives no information about where the intruding aircraft is located. The fact that tau has slipped below forty seconds implies the intruder is close enough to be seen if conditions are good. The choice of the value of tau to activate the traffic alert depends on the type of aircraft. A highly maneuverable aircraft may choose a lower value of tau because the escape maneuver can be quite abrupt. For air-transport aircraft, however, abrupt maneuvers are not possible.

The more sophisticated TCAS installation, which represents what is found in air-transport-category aircraft, has a two-dimensional display of the proximate traffic. This aids the crew in spotting the intruder aircraft and taking evasive action. Air-transport-category aircraft always fly under air-traffic control, and any evasive maneuver would violate their ATC instructions. If there is a danger of collision, an evasive maneuver can change the separation between "own aircraft" and other aircraft. In heavy air traffic, a sudden, unexpected maneuver could cause another aircraft to receive a traffic advisory and make an adjustment. This is the classic example of the "domino effect." If aircraft are neatly separated, one that suddenly moves out of its assigned place will cause the separation to another aircraft to decrease, which produces a traffic advisory that causes the second aircraft to take evasive action, and so on. Because of this, it is vital that TCAS never provide false alarms.

TCAS was successful while other transponder-based, collision-avoidance systems were not—for three reasons. First, TCAS uses a directional antenna. The added information this provides allows for much more accurate decisions to be made about possible intruders. Second, TCAS operates with a more sophisticated transponder, mode S, than the old ATCRBS relic from World War II. Finally, the computing power available in the early 1980s was a thousand times greater than that available in the early 1950s, when electronic collision avoidance was first contemplated. The typical TCAS computer has multiple processors, including at least one 80486 or Pentium-class computer.

TCAS is required on all air-transport-category aircraft and all larger cargo aircraft. Experiments were conducted on the use of TCAS for reducing the separation of aircraft out of the range of ground-based radar. Aircraft off shore by more than about 100 nautical miles and not close to an island, which represents most of the oceans of the world, and in the polar regions are out of range of ground-based radar. Air-traffic control ensures safe separation in these areas by using position fixes from the aircrew. Most of these position fixes are obtained via HF radio because the aircraft are also out of VHF radio range. Because of the congestion on the HF radio channels and because of atmospheric noise, several minutes might elapse between an initial call from an aircraft and the air-traffic controller's receipt of the aircraft's correct position. Because of this problem, the separation "in trail" is typically 60 nautical miles. *In trail* refers to aircraft on essentially the same course. Lateral separation can be considerably less, perhaps 20 nautical miles, and vertical separation can be about four thousand feet.

Such large in-trail separations reduce the capacity of some of the more popular transoceanic routes. The use of TCAS can reduce in-trail separations considerably. Most TCAS units can display the position of proximate aircraft to a maximum range of 20 nautical miles. Although not all aircraft are required to use TCAS, in practice there are no aircraft in transoceanic flight without it. The use of TCAS for maintaining safe separation can reduce the in-trail separation to 30 nautical miles or less, which effectively doubles the airways' capacity.

Controlled Flight into Terrain

Controlled flight into terrain, or CFIT (pronounced see-fit), occurs when an aircraft is flying perfectly fine but crashes into the ground or a structure. *Controlled flight* means the aircraft is not spinning out of control, exceeding any maximum rating, or seriously damaged. A typical CFIT accident might involve a situation in which a crew, for whatever reason, believes they can descend for landing when they are actually too far out and instead descend into the side of a mountain. More common than most people realize, this scenario characterizes the type of accident that caused the FAA to push for the development of a system called "ground-proximity warning system."

The ground-proximity warning system, GPWS, or sometimes called "ground prox," is an adjunct to the radar altimeter. There are three types of altitude used in aircraft. The first is altitude above mean sea level, or MSL; the second is pressure altitude; and the third is above ground level, or AGL. *Mean sea level altitude* is used for aircraft at terminal areas and in low-altitude flight. *Pressure altitude* is used for high-altitude flight. Both MSL and pressure altitudes are determined by barometric measurements. *Above ground level altitude* is determined by the radar altimeter.

The radar altimeter is a very precise, short-range radar. The altimeter has a maximum range of about two thousand feet and can determine altitude to within about ten feet. The radar altimeter is used for the higher categories—II and III—of precision instrument landings.

The ground-proximity warning system uses the radar altimeter to simply gather data about the distance to the ground. For example, if the distance to the ground is rapidly decreasing while the aircraft is not descending according to the barometric altimeter, this is a likely indication that the aircraft is flying into a mountain. This is just one possible scenario, and information is fed to the ground-proximity warning system from a number of other aircraft systems.

As with any warning system, false alarms are to be avoided as much as possible. Consider the example given where the above-ground altitude is decreasing. However, in this scenario let us assume the aircraft is properly oriented on an instrument landing beam. The situation is simple: The aircraft is making a landing, and the decreasing altitude is normal. The GPWS receives information from the instrument-landing system on the aircraft. Information concerning the status of the flaps and landing gear is also available. The GPWS computer realizes the ground is approaching, but because the aircraft is on a valid glide path and is configured for landing, no warning is issued. On the other hand, if the aircraft is below the glide path, a warning is in order.

The ground-proximity warning system is one of the first airborne systems to use a voice announcement. One of the problems with warning systems is that when there are too many warnings, the lights, gongs, bells, whistles, and klaxons become an annoyance, and aircrew members begin to ignore them. Before the ground-proximity, voice warning system, aural alarms in larger aircraft included the marker-beacon tones, a horn warning of impending stall, gear-up warnings, autopilot disengage warning, altitude-alert tones, and so on. With the exception of the marker beacon, which was the same for all aircraft, the warning devices were not standardized.

A story was related to this author by a friend who was a control-tower operator at a municipal airport in New Jersey. The tower had cleared a small, retractable-gear aircraft for landing and then noticed the aircraft's gear was still up. The tower called the aircraft on approach and warned the pilot that the gear was still up. The pilot responded by radio, and, as he was calling the tower, the controller could hear the sound of the gear-up warning horn in the background. The pilot called the tower and said, "Say again, please. I did not hear you. There is some sort of loud honking in here!" Fortunately, the tower operator told the aircraft to "go around" and explained what the "loud honking" was. This story shows that if an alarm is not easily identifiable, the aircrew will spend precious time trying to determine what the alarm means rather than correcting a dangerous situation. The GPWS's aural warning is the announcement, "Terrain! Terrain!" If the situation persists, the announcement is "Pull up! Pull up!"

A similar system is used for the TCAS warnings. At the time TCAS and the ground-proximity warning system were being designed, electronically synthesizing human speech was a rather simple matter. If these systems had been designed before the advent of microprocessors, making a voice alarm would not have been so simple, and most likely there would have been another gong or honk in the cockpit.

A more recent system to reduce the chance of controlled flight into terrain is the enhanced, ground-proximity warning system, or EGPWS. This system uses the aircraft's position as determined by GPS and a stored database showing mountain peaks, tall towers, and so on. The FCC has been assembling a database of the location of radio towers, and these data are used to locate possible collision obstacles. In addition to having an aural warning system, the EGPWS also has a graphics display that shows the location of high points in the area in which an aircraft is flying. The EGPWS has been approved by the FAA and given a new name: the "terrain-awareness warning system," or TAWS, and will become mandatory in certain aircraft.

When Accidents Happen

Accidents happen. Government agencies around the world have taken every precaution to prevent accidents, but when they occur, it is essential to determine the cause as soon as the rescue or recovery has been completed. Two systems provide some of the most effective tools to analyze accidents: the cockpit voice recorder (CVR) and the flight-data recorder (FDR).

The cockpit voice recorder is relatively easy to understand. The earliest units were nothing more than rugged tape recorders. The most important requirement for the CVR is that it has to survive a crash. The recorder did not have to survive, but the tape did. The tape-drive mechanism with the tape was placed in a very strong housing that would survive water, fire, and impact. The voice recorder would record normal conversations from the captain and first officer's microphones, and a third or even fourth microphone was mounted in the cockpit so that casual conversation between the flight crew could be monitored. Even though the device is called a voice recorder, much of what it records is not voices yet is still valuable in accident investigations. Sounds such as explosions, engine noise, wind noise, crew alarms, and gunshots are very valuable to analysis.

The modern version of the CVR uses an electronic storage medium with memory chips of the type that retain data without applied power. One of the problems with recording tape is that it records over and over again and eventually begins to deteriorate. The solid-state recorder provides the same, high-quality recording regardless of how many times the memory has been written. Like the tape drive, the memory is sealed in a very strong housing that will survive water, fire, and impact. The early cockpit voice recorders recorded for thirty minutes before overwriting, while the newer recorders can handle up to two hours.

The flight-data recorder records various data such as altitude, airspeed, engine RPMs, and so on. The earliest recorders recorded the data on strips of metal foil by scratching a trace on the foil. Because some tapes were very hard to read, later recorders began to use digital-storage techniques similar to those found in the voice recorder. The latest flight-data recorders can record from 5 to 300 channels for twenty-five hours with superb fidelity.

It is important that these recorders be found after a crash. For whatever reason, news reporters always refer to the recorders as "black boxes." It is true that virtually every piece of electronic equipment aboard an aircraft is housed in a black box—except the recorders, which are orange. The bright-orange color is a deliberate design feature so that the recorders can be easily spotted among all the black boxes in an aircraft. The second aid to finding the boxes is the "pinger." This is an electronic device that emits an ultrasonic sound that can be picked up by a sonarlike device at great distances on land and in water. Thanks to international standards, all pingers emit exactly the same sound, so that the same locating devices can be used with any brand of "orange box."

Sometimes it is necessary to locate a downed aircraft in a much larger area than a pinger can service. An electronic locating system that is used in nearly all aircraft, from the largest to the smallest, is the emergency locator transmitter, or ELT. All ELTs are automatically activated by an impact switch. If an aircraft does not land with a great impact, it may be assumed that injuries are not very serious and the ELT may be manually activated. Some transmitters are activated by contact with water. The most sophisticated ELTs are actually ejected away from the aircraft and then activated. This is done so that if a fire breaks out, the locator transmitter is away from the aircraft and will survive.

The standard emergency-locator beacon operates on the international emergency frequencies of 121.5 MHz and 243 MHz. Towers, pilots, airports, and others monitor these emergency frequencies and will recognize the distinctive wail of an emergency beacon. The ground-to-ground range of the beacon is only a few miles. From ground to air, the range is on the order of fifty miles. Rescuers must find the beacon by first hearing its transmission and then using direction-finding techniques. If a crash has occurred in a desolate place, quite some time could elapse before rescue efforts start.

A more sophisticated emergency location system uses the search-and-rescue satellite (SARSAT) system. The advantage of this system is that the SARSAT can receive an emergency-beacon's signal anywhere in the world within a few minutes. The newest rescue beacons operate in conjunction with the GPS navigation system and transmit a very accurate position. Therefore, within minutes after a crash, rescue teams know the position of the downed aircraft to within a few meters. Many nonrescue satellites, such as the GOES series of weather satellites, monitor the 406-MHz emergency frequency. The intent is to have at least one satellite within range of every location on Earth. Therefore, the response to a SARSAT beacon will be immediate, regardless of where the crash took place.

A New Look: The Glass Cockpit

The goal of every navigation system is to present the aircrew a picture of the aircraft's situation within its surroundings. Even the very early navigation indicators attempted to "point" to a navigation station or to an airway. The first gyro instruments were painted to look like the sky and ground on the artificial horizon with a silhouette of an aircraft. The two major instruments to emerge were the horizontal situation indicator, or HSI, and the attitude direction indicator, or ADI.

The HSI is the major en-route instrument and displays the aircraft's heading, the selected course, and the aircraft's position relative to the desired course. In the center of the indicator is the outline of an aircraft, which gives the pilot a picture of "own" aircraft and the desired course.

The ADI is a sophisticated version of the artificial horizon and is the main indicator used for landing. The roll-and-pitch angles are visible on this indicator as well as course information, both vertical and horizontal, during an instrument landing.

The Collins Radio Company designed and introduced to the industry the first in a line of pictorial navigation instruments. In the process of some instrument maneuvers, such as approach to landing, the flight crew must monitor a number of instruments to make a smooth and safe approach. The artificial horizon, along with the turn-and-bank indicator, ensure smooth, co-ordinated turns. The heading indicator provides guidance to intercept the instrument landing system's beam. When on the beam, the flight crew use the glide slope and localizer indicators to remain on the landing beam, while the artificial horizon helps the crew keep the wings level and maintain a reasonable angle of attack. The difficulty is in tracking several instruments simultaneously. For a visual landing, a look out of the window provides a view of the entire situation. What would reduce the workload would be a replacement for the actual view from the flight deck synthesized by the electronic systems aboard the aircraft. In 1953 Collins Radio introduced a device to do just that: a *flight director.*

The flight director put together the artificial horizon, the turn coordinator, the glide slope and localizer pointers, and other indicators and warning lamps. In the center of the indicator's glass window was the outline of an aircraft. The background of the indicator was the outline of a runway in perspective. With a little bit of imagination a pilot could visualize an aircraft making an approach to a runway.

These indicators were electromechanical masterpieces, or, more accurately, electromechanical nightmares. They were extremely expensive and, because of all the moving parts, were often in need of repair and adjustment. The idea to replace these instruments with an all-electronic system had been around for quite a while. In the mid-1970s, advances in small, cathode-ray tubes moved the idea closer to reality. The cathode-ray tube (CRT) was already in use in a number of aircraft as the weather radar indicator. The first airborne weather radar, produced by Bendix in 1953, was a monstrosity that used a large indicator that would not fit behind the instrument panel. The indicator was stood up on end near the center console. The display had no storage, and a long-persistence phosphor was used on the CRT. The only way to see the dim display was to view it through a hood that would block out the ambient light. When integrated-circuit memory chips became available in the 1970s, the image could be stored electronically and not on the CRT face. This greatly improved the image, but the displays were still a monochrome green.

The CRT is the device that creates a television picture and most computer displays. There were a number of problems with using a CRT in an aircraft. The first is that the tube is somewhat fragile, particularly the color tubes. Another problem is that the tube is long, and the space behind the instrument panel is not large. Another problem was that CRTs are not extremely bright. The majority of CRTs were initially for television receivers and, later, computer monitors. Both of these devices do not operate in full sunlight. Finally, the CRT is a vacuum tube, which requires high operating voltages. The vacuum-tube era had been over for a long time, and high voltage was not readily available. The capability of the CRT to generate pictorial displays so outweighed its problems that avionics manufacturers found solutions for the problems and made use of the CRT.

One of the first CRTs considered for aircraft use was the Sony Trinitron. This tube was in high-volume production for portable, color-television receivers and was a natural for a small-aircraft indicator. The Trinitron had an improved design that made the tube inherently rugged.

High voltage in a CRT cannot be avoided, so the manufacturers designed the required power supplies. Power transistors can safely generate high voltage and had been used for this for years in television sets. It was necessary for the high-voltage power supply to be rugged and able to operate from the usual aircraft, power-supply voltages.

The length of the CRT was shortened somewhat by using electromagnetic deflection, as in a television receiver. The greater the electron beam that can be deflected, the shorter the tube can be. Reducing the length of the CRT was a goal of television manufacturers, so the newest tubes were reasonably short, particularly those in portable television receivers. The first aircraft CRT displays were about six inches diagonally measured and were based on television receiver designs.

Obtaining brightness is a matter of using high-beam currents and bright phosphors on the face of the tube. This made for an expensive, short-lived tube, but when compared to the complex electromechanical instruments it replaced, the advantages of the CRT outstripped its disadvantages.

One of the first applications of a small color display was weather radar.

The radar already used a CRT display, so it was the obvious place to put the first color display in the cockpit. One of the most significant advantages of the CRT is that it can display anything. The manufacturers of weather radar decided that they could offer features such as a checklist on the radar display. The real advantage of the CRT would come into play when the electromechanical indicators could be replaced with the CRT display.

The first CRT indicators replaced the HSI and the ADI. But why stop there? Nothing prevented all of the electromechanical indicators from appearing on CRTs. A new system that relied on CRT displays was the engine-instrumentation and crew-alerting system, or EICAS, pronounced eye-kass. This system is more than a CRT replacement for the old electromechanical indicators called "steam gauges." The EICAS was capable of analyzing engine data and reporting when a particular parameter had fallen out of the normal range. A system of this type reduced the aircrew's workload. This is particularly important on large aircraft, where the role of the flight engineer could be eliminated.

The use of color liquid crystal displays (LCDs) made portable computers practicable. The first "portable" computers used a CRT display and were doomed to failure for that reason alone. The color LCD was refined to a point where bright, high-resolution displays became affordable. The LCD would solve the depth problem and save considerable power, too. But the cockpit environment is a very demanding one, and at the time of this writing, CRTs and LCDs are both found in glass cockpits. The LCD will emerge the victor as soon as the problems are solved.

The Head-Up Display

The most difficult maneuver in flying is making an instrument approach and landing. All instrument landings end with a visual touchdown. As soon as the runway is seen, the instrument landing is over, and the aircraft is landed using visual means. This is best done with an autopilot and two pilots. The attention of one of the two pilots is focused on the inside of the aircraft, while the other pilot focuses on spotting the runway. One of the problems associated with both focusing one's attention on the inside of the aircraft and attempting to spot the runway is the need to constantly refocus the eyes from the instrument panel to the outside.

Category III landings require the use of an autopilot so that the pilots do not have to constantly refocus. A device borrowed from the military to permit category III instrument landings without an autoland system is the head-up display. This system produces an image of the necessary instruments on a glass plate in front of the pilot, in a way that is similar to the reflection one would see on the windshield of an automobile. The difference between the head-up display and a reflection on a windshield is that the head-up display's image is focused at infinity. Therefore, the pilot's eyes are focused at infinity, where the simultaneous images of the instruments and the terrain are seen. The original application was for military fighters, so that the pilot could focus on the target ahead while the fire-control indicators are still perfectly visible. In civilian applications, the glass cockpit can project any instrument on the head-up display in the same fashion.

Chapter 8

The Electronic Airplane
Learning to Fly without an Aircraft

During World War II, when pilots were needed in huge numbers, many aspiring pilots were introduced to a machine called the "Link trainer." The trainer was a replica of an airplane cockpit with a complete instrument panel. Although the device looked like an arcade ride for a child, it was the first aircraft simulator. This device taught pilots how to hold altitude, make a turn, and follow a simple course. Although of limited capability, the link trainer improved pilot training.

Crude by modern standards, the link trainer was nevertheless quite an accomplishment for its time. The trainer would roll, pitch, and yaw in response to the stick-and-rudder pedals and could simulate climbs up to fourteen hundred feet per minute. The controls would "mush," or become less responsive, for low airspeeds. The trainer was equipped with a turn indicator, airspeed indicator, turn coordinator, vertical speed indicator, altimeter, tachometer, clock, compass, directional gyro, and an artificial horizon. The only radio navigation system in operation at the time the trainer was introduced was the AN range. There were recordings of AN range signals corresponding to various on-course and off-course situations. Since the link trainer predated magnetic tape recorders, the recordings were on magnetic wire recorders.

With the exception of the clock, the instruments in the trainer were all simulated. To this day, the best method of simulating an instrument is to use the actual airborne instrument and excite it with a signal. As an example, the tachometer in the link trainer is connected to an electric motor. The throttle in the trainer controls the speed of the electric motor using a conventional rheostat. Up to a point, this provides a realistic simulation. Pushing the throttle forward causes an increase in the electric motor speed but with a delay, just as would occur with an aircraft engine. Likewise, closing the throttle causes a decrease in engine RPMs, again with a delay. Connecting a second rheostat to the pitch axis of the trainer causes the engine RPMs to decrease when the simulator is pitched up and increase when the simulator is pitched down.

Simulating other instruments is not as simple. Consider the compass and directional gyro (DG). Since the trainer does not actually turn, neither instrument can be used unmodified, which was possible with the tachometer. Therefore, the magnet in the compass and the gyro in the DG are disabled or removed. Both instruments provide the same heading information. The instruments are highly modified by connecting the compass rose in each instrument to a special motor called a "selsyn," which is short for "self-synchronizing," or simply "synchro." In its simplest application, two synchros are connected with wires, and the shaft of the two motors is synchronized. One synchro is called the transmitter and the other, the receiver. If the shaft of the transmitter is turned to a specific position, the receiver will mimic that position.

In the simulated DG, a conventional electric motor drives a transmitter synchro through a gear reduction. This motor is connected to the rudder pedals, so the motor turns in one direction when the pedals are pressed to the left and the reverse rotation when the pedals are pressed to the right. Along with the ailerons, the rudder causes the aircraft to turn; the more the rudder pedals are pressed, the greater the rate of turn. As long as the simulator is in a turn, the electric motor turns, and the DG and the compass indicators continue to rotate. Once the simulator is no longer turning, the DG and compass stop. The rate of turn is proportional to the speed of the gear reduction motor. Since the motor speed is proportional to the voltage applied to the motor from the rudder pedals, that signal can also be used to drive the rate-of-turn indicator.

The DG and the compass do not always provide the same information. When an aircraft is flying straight and level, the compass and the DG should read the same. When the aircraft is in a turn, the compass exhibits errors, which depend on the heading and the bank angle of the aircraft. Early trainers simulated this error by using heading information from the DG along with the roll information and inserting errors in the simulated compass by using mechanical differencing circuits.

The link trainer is a fascinating device because the entire simulation was done without a computer. Before the advent of the first analog computers and later digital computers in the mid-and late-1940s respectively, mechanical computing had reached a high level of sophistication. Complex equations were solved with mechanical behemoths called "differential analyzers." Some of these mechanical wonders were used for a few years after digital computers became available. Not long after analog and digital computers became available, flight simulators made use of the new technology. However, synchros, electric motors, and gear boxes are still found in many aircraft, driving indicators, and other applications. Only with the introduction of electronic displays did these mechanical components begin to disappear.

The link trainer was the first of what is called a *simulator* today. Many people have the false notion that a simulator is an inferior, low-cost method of training. Nothing could be further from the truth. First, for smaller aircraft, a high-performance simulator can cost more than the actual aircraft. In most cases the operating costs of a simulator are lower than those for an aircraft, particularly if the simulator is operated for twenty or more hours per day.

The modern simulator has full motion, visuals, and realistic sound. The simulated cockpit is in a sealed cab with no visual or audio reference to the outside. The cab is capable of extensive motion under the control of hydraulic pistons. The simulated aircraft has windows, but the "view from the window" is a projection screen. These projection screens are connected to the cab, so the motion of the cab affects the view from the window.

Some very clever techniques simulate motion. For example, when the throttles are advanced, the simulator cab tilts back, but the out-the-window visual does not change pitch. What happens, however, is that the runway begins to pass under the aircraft in accelerated motion, and the airspeed indicator begins to increase. To the crew, inside the cab, being pressed back into their seats is not due to the tipping (which they cannot see) but rather to acceleration. When the aircraft reaches rotation velocity, the pilot pulls back on the controls, and the simulator cab tips more, but the visuals now show a nose-up attitude. The crew experiences an increased force, which would be correct for an aircraft not only accelerating but now also pitched up in a climb. When this is coupled with the sound of wheels rumbling on the runway, the vibration of the takeoff, and the sound of the landing gear stowing in the fuselage, the feeling of a takeoff is very real.

The instruments (such as the attitude direction indicator) also show the nose-up attitude, and the altimeter begins to increase. The gear is raised, and in the simulator, from a loudspeaker under the cab, the crew can hear the normal sound of the nose gear being raised. The sound from the loudspeaker vibrates the floor of the cab as the gear is stowed, and the increasing airspeed can be heard and felt from the same loudspeaker. The noise of various systems operating in the aircraft are derived from actual sounds in the aircraft. The windshield wiper in one specific aircraft model has an unusually obnoxious noise in the flight deck, and this sound is faithfully reproduced.

When so many small details are reproduced, the feeling of actual flight in an airborne aircraft is quite realistic, and the simulator is an effective learning tool. What can also be done with a simulator but cannot be done with actual aircraft is the simulation of unusual and emergency conditions. Crew can be trained (and their performance evaluated) in how to handle engine out, total loss of power, icing, and so on. Simulation extends even beyond the flight deck. Simulated cabins are used to train cabin attendants for emergency situations, such as ditching and fire.

Another very effective application of simulators is in investigating crashes. Information from the flight data recorder of an aircraft involved in a mishap is used to recreate the displays of the flight deck. A crew is placed in the simulator and on the flight path just before the end of the flight is replicated from the data downloaded from the flight data recorder. By studying the actions of the crew in the simulator, investigators gain valuable insight into the probable causes of the crash. Simulation has also found applications beyond the aviation business. Simulators are available for boats, army tanks, automobiles, and subway cars. There is even a simulator to train surgeons in performing surgery.

Collision Avoidance Revisited

The major problem with TCAS was its cost. A TCAS system, fully installed in a large aircraft, costs as much as a factory-new, high-performance, single-engine aircraft. Mandating TCAS for air-transport aircraft was successful because an expensive piece of electronic equipment aboard an already expensive aircraft is more palatable than installing a system that costs as much as the entire aircraft.

The GPS system provides the core of an inexpensive, collision-avoidance system. If an aircraft is equipped with a GPS receiver, that aircraft knows its latitude, longitude, and altitude. If every aircraft simply transmitted its position as determined by the onboard GPS receiver using a short-range, radio-data link, other aircraft in the vicinity could receive the positions of all aircraft within range of the data link. The system that operates on these principles is called "automatic, dependent surveillance, broadcast mode," or ADS-B. Unlike TCAS, an aircraft would have to have only a low-powered transmitter and a receiver. TCAS required a mode-S transponder, an interrogation transmitter, and a directional antenna.

In order for ADS-B to be successful, all aircraft must be equipped with ADS-B equipment. In the case of TCAS, a TCAS-equipped aircraft could be protected from an aircraft that was only transponder equipped. However, FAA regulations require transponders in practically all airspace. Forcing the nearly universal use of a transponder met great resistance. In an air-transport-category aircraft, the cost of a transponder is negligible—not so with a general-aviation aircraft. If the required device benefits the GA-aircraft owner, the likelihood of the owner installing the equipment is greater than if it offers no benefit. When the FAA mandated the use of transponders, it was primarily the air-traffic controllers who benefited; many aircraft owners viewed the device as an invasion of privacy. ADS-B requires the installation of a GPS receiver as well as a data transmitter and receiver. The GPS is a welcomed device in an aircraft because it enhances the aircraft's navigation capability. As long as the data link is not expensive, there is hope that general-aviation owners can be convinced to install ADS-B.

Redefining Air-Traffic Control

The concept behind air-traffic control, which was introduced in the 1920s and discussed in previous chapters, was to place aircraft on airways and assign the responsibility of ensuring safe separation to air-traffic controllers. Early controllers determined position by reports from aircrews and later from radar data. In terms of energy efficiency, time en route, and comfort, the federal airways may not be the best choice. Pilots may ask for different cruising altitudes for a less turbulent ride, but other than that, there is little choice in an instrument-flight route. Ideally, an aircraft would like to leave when the pilot wishes and fly the most direct route at an altitude that provides the greatest comfort or fuel economy.

The concept of federal airways was based on navigation aids—first, the light beacons, and later, radio navigation. What happens in this situation is that all en-route aircraft are funneled onto airways while most of the air space is empty. Flying a specific course was a part of the navigation systems in use such as the AN range, nondirectional beacons, and VHF omnirange. The later, long-distance navigation aids (such as the LORANs, Omega and, most important, GPS) have virtually no limitations on the choice of course. The obvious course to fly with GPS is a straight line from departure airport to destination airport. Because the Earth is not a flat surface, the straight line is actually curved and is called a "great circle" route. Because a straight line is the shortest distance between two points on a flat surface, the great circle is the shortest distance between two points on the surface of a sphere. If the rules of instrument flight were changed so that any aircraft could fly any course desired, the first difference we would note would be that the en-route aircraft would be spread out over a vast area rather than concentrated on airways. Therefore, on average, the separation between aircraft would increase.

Perhaps one of the first thoughts one has when contemplating such a situation is, how would air-traffic controllers keep track of a great number of aircraft going every which way? The answer is simple: There would be no air-traffic controllers. The aircraft have set their own courses. Eventually there would be a "conflict," that is, two aircraft that would occupy the same airspace at the same time, and this must, of course, be managed. Air-traffic control then would become air-traffic management, or ATM.

There must be a method of predicting—and thus eliminating—these potential conflicts in sufficient time to alter the course of one or more of the aircraft. A computer could easily handle this job if it could be fed the position of every aircraft in a managed area. As a "conflict probe," the computer constantly predicts future conflicts from the available data. In addition to the current position of the aircraft, the ADS-B broadcast includes "intent," which is the estimated course of the aircraft. This aids in predicting a future conflict. Called "free flight," this technique will become the standard method of air-traffic management in the first decade of the twenty-first century.

The conflict probe is not a collision-avoidance device. TCAS, for example, monitored proximate aircraft within only a 20-nautical-mile radius of "own aircraft." Warnings were issued at 40 seconds and 25 seconds. The conflict probe alerts air-traffic managers of potential conflicts many minutes in advance; flight planning can then avoid a collision rather than perform an evasive maneuver. Aircraft flying in opposite directions, each doing 500 knots, would approach at 1,000 knots. Aircraft that are 100 nautical miles apart will meet in six minutes. Thus air-traffic management should be aware of the conflict and take corrective action. The positions and intents of two aircraft separated by 100 nautical miles should be relayed to an air-traffic manager, who will contact both aircraft and resolve the conflict. The two aircraft will not necessarily be in radio range of an air-traffic manager or the conflicting air-craft. Thus remote communications sites will relay the information between the air-traffic manager and the two aircraft. This requires a network of all ground-communications facilities and all airborne aircraft and air-traffic man-agers. This network is called *aeronautical telecommunications network,* or ATN.

In the "information age," more information valuable for flight planning is available on the ground than in the air. For example, a well-equipped air-craft would have an onboard weather radar that would provide the air crew with weather in the immediate vicinity out to about one hundred nautical miles. Internet users would be able to get the same weather information as well as information for the entire en-route trip and the destination. The on-board weather radar provides a high-definition and timely picture of the weather in the immediate vicinity, but Internet users have access to consider-ably more data of all types. There has been a lack of data available to en-route aircraft, but a number of proposals and programs will help to change this sit-uation. The first onboard data will include clearance and flight-plan data. En route and weather and weather forecasts for all destinations will also become available. A number of data links will be available in the modern aircraft.

A New Source of Interference: Portable Electronics Carried Aboard

The role of electronics in the evolution of flight is not all positive. The aircraft of the 1950s were already bristling with electronic equipment, which included a number of radio receivers. Transistors, which were invented by Bell Labo-ratories in 1947, had progressed to a point where they were finding applica-tions in electronics. One of the first applications outside of the telephone company and the military was in portable radios. In 1954 Regency, a division of Texas Instruments, made the first, portable, transistorized radio receiver.

The Regency portable radio was by no means the first portable radio ever marketed. Portable radio receivers had been sold since the 1930s. How-ever, these older portable radios, with a few exceptions, were very large and heavy and had batteries that lasted for only 20 or 30 hours. No one would carry one of these monsters on board an aircraft. The new portable radio, however, was small, weighed only slightly more than a pound, and had a battery life of hundreds of hours. This radio was an excellent source of diversion during a long trip.

All radio receivers radiate a small amount of radio energy. This was discussed relative to the AN/APR-4 airborne surveillance receiver, which gave its own position away. In the early days of consumer radio, no effort was taken to minimize the amount of radiation, and some receivers were responsible for causing considerable interference to other receivers. This prompted the FCC to limit how much a radio receiver can radiate by requiring designs to be certified. When avionics are installed in aircraft, the installation is checked for interference by operating all of the avionics in as many different modes of operation as practicable, and the effects of interference are examined. If a new piece of equipment is installed, the process is repeated. This exercise determines *electromagnetic compatibility*, EMC, and is a very important part of modern aircraft design.

After the aircraft is carefully checked for potential interference, the door is opened to hundreds of passengers carrying all sorts of electronic equipment that they intend to use during the flight. So much for the value of the careful EMC testing.

In the late 1950s and early 1960s, the effects of interference were being seen in the navigation equipment of air-transport aircraft. In 1958 the FAA requested that the RTCA form a special committee to investigate the problem and prepare a report. The conclusion of the report was that a potential problem existed but that it was not severe enough to warrant prohibiting the portable electronic devices on aircraft. The FAA charged the airlines with ensuring the safety of air-transport aircraft, and the airlines subsequently restricted the use of portable electronic devices to the en-route phases of flight.

When the problem reappeared in 1987, the FAA again requested a special committee of the RTCA. During this time there was a virtual explosion of portable electronic equipment including tape players, handheld games, calculators, and video cameras. The report was the same, however: There seems to be no immediate problem. Virtually nothing had changed.

In 1993 the FAA requested for a third time that the RTCA investigate the problem. This time the culprits were the new, portable digital devices, most notably the laptop or notebook computers. The results were the same, except the warning was sterner. The problem is not going away, the report warned. The RTCA report also received significant press coverage, and the problem went as far as congressional hearings in July, 2000.

The problem has not subsided, so testing continues. At the time of this writing, a test that the Civil Aviation Administration of the United Kingdom performed in early 2000 determined that cellular telephones produce radiation levels above those for which avionics are tested and certified. Although a rare problem and not yet the cause of any serious mishap, reports of incidents have appeared. In October, 2000, an aircraft carrying the German foreign minister had to make a landing without an unspecified system in operation, which was attributed to onboard cellular telephone use.

Flying by Wire

Ever since the very first powered flight, there has been a physical connection between the pilot and the aircraft. In the Wright flyer, Orville controlled the aircraft by shifting his weight. Later aircraft were controlled by a system of levers and pedals or stick and rudder. This intimate connection between pilot and aircraft helped in the safe control of the aircraft. A direct mechanical connection permitted the feeding back of the forces on the control surfaces, which help prevent overcontrol or placing the aircraft in dangerous situations, such as stalls.

As aircraft grew in size, the forces required to control them were beyond the capability of a human operator. To handle these larger aircraft, assist systems were employed. These were very similar to the power assist in cars for so-called power steering and power brakes. In these systems a direct mechanical connection still exists, but hydraulic cylinders provide additional force to control the front wheels, brakes, or control surfaces. If the power-assist system should fail in a car, the vehicle could still be stopped with the brake pedal, although the required force would be much greater. Likewise, the car could be steered if the power assist should fail, but the steering would become difficult. However, it would not be humanly possible to control a large aircraft with no power assist; these systems, therefore, are critical to safe flight.

In power assist, a direct mechanical connection remains with the object to be controlled. This ensures a very important characteristic of these systems. In the case of a car with power steering, if the steering wheel were turned by, say, one-half a turn, the front wheels would turn by a certain amount. The amount of motion of the front wheels for half a turn of the steering wheel is the same regardless of whether the power assist is working. The same is true of a power-brake system. The harder the operator presses the brake pedal, the more stopping power is directed to the wheels. Astute readers will immediately counter that this is not true of the antilock braking system. This is a perfect example of the difference between a simple assist system and an automatic control system.

In the antilock brake system, the rotation of the wheels is sensed and fed to a computer. Should one of the four wheels turn at a slower rate than the other three, the system assumes that the wheel is skidding. It then reduces the braking power until that wheel begins to turn at the same rate as the others. Let us consider what happens when a panicked driver attempts to stop on pure ice. The antilock braking system will reduce the stopping power to all four wheels, and most likely the driver—sensing that the vehicle is not stopping—will increase the pedal pressure. This is actually the wrong thing to do, and the driver's added pressure is to no avail. The antilock braking system receives input from the driver—the depressed brake pedal—but the system takes command of braking the vehicle. In a conventional power-assist braking system, the driver's increased brake pressure will cause all four wheels to lock up on the ice, putting the car into an uncontrolled, four-wheel skid.

A similar thing could be done with the power steering of a car. Let us assume that a car has a sensor that can determine when the car is in a spin. If a driver makes a turn on a slick road at such a rate that the rear wheels begin to drift to the outside of the turn, the power steering will stop the skid by reducing the amount the front wheels have turned. The driver will sense that he or she has lost the connection between the steering wheel and the front wheels, which is exactly what has happened.

Like a car, an aircraft has some unpleasant maneuvers. A control system could be used in an aircraft that will take badly performed maneuvers and undo the mistakes so that the aircraft performs a textbook-perfect maneuver. In general, pilots are not poor performers, but some aircraft behave badly. This is not a bad thing. Let us consider, for a moment, the behavior of aircraft, particularly what is called *stability*. If the pilot lets go of the controls of a well-behaved aircraft, the aircraft will assume straight-and-level flight, even if the controls are let go during a turn.

The same is true of a good car. Drivers know that if they let go of the steering wheel during a turn, the vehicle will cease turning and assume a relatively straight path. Another way of looking at this situation is that a stable aircraft or car has to be forced to turn, as its tendency is to go in a straight line. This characteristic is fine for passenger cars or air-transport aircraft. However, for a race car or a fighter aircraft, turning may be more important than moving in a straight line. Race cars and fighter aircraft are made to be less stable and easier to turn and maneuver. This makes these vehicles difficult to drive or fly. If we were to take the stability issue to the extreme in an aircraft, a very unstable but very maneuverable aircraft would be beyond the capability of a human to fly.

What if a computer were used to overcome the difficulties of the unstable aircraft? What if the computer took inputs from the human pilot and caused the aircraft to do what the pilot desired? This would be similar to the antilock braking system, where the human driver put maximum pressure on the brake pedal, but the braking system provided only partial braking power to some wheels so that the vehicle would not skid out of control. In fact, the system provided the maximum braking power for the situation.

This type of system is called "fly by wire." This type of system is mentioned briefly in a previous chapter relative to the automatic flight-control system. In *fly by wire,* the input from the pilot is via an electrical device. There is no mechanical linkage between the control device in the cockpit and the control surfaces. The control device provides signals to the flight-control computer, which receives these data as the pilot's intent, decides exactly which control surfaces will move and by how much. When used with an aircraft that is inherently unstable, this system is called a *stability augmentation system.*

This concept opens up some rather interesting possibilities. One of the first was to reduce the size of the crew controls in an aircraft. Historically, a stick and rudder pedals or a yoke (similar to a steering wheel in a car) with rudder pedals were used to control an aircraft. Also, these controls moved through quite a distance because they were originally direct-connected to the control surfaces. The amount of motion was on the order of one-half meter or more to provide sufficient mechanical advantage to prevent the force from being excessive. The larger the aircraft, the greater the motion. The range of motion is so great that large aircraft have special seats for the crew. These seats allow the yoke to be pulled between the legs up to the pilot's chest. When assist systems were added to large aircraft, theoretically the range of motion could be reduced considerably, but it was not. This was done so that the motions familiar to pilots would remain unchanged.

In aircraft where cockpits are small and maneuvers occur frequently, large controls and broad movements are highly undesirable. In the modern fighter aircraft, the stick and rudder were replaced by a small control, called the side stick controller, that uses the wrist motion of only one hand. This controller sends signals to the flight-control computer, which then controls the aircraft.

This total disconnect of the pilot from the control surfaces had both good and bad points. On the good side, learning to fly an aircraft with the new side stick controller was easy, and pilots were mastering the technique in a short time. On the bad side, the aircraft was like the proverbial genie in a bottle: Your wish is my command. If the pilot quickly pushed the controller to one extreme or another, the aircraft would oblige the pilot by performing a violent maneuver. At times, to the military pilot, a violent maneuver means the difference between life and death. But in older aircraft, the resistance on the controls reminded the pilot that the aircraft was being pushed to its limits. To make extreme maneuvers, the pilot had to exert considerable force. Then the pilot could feel that the aircraft was approaching its limits.

In order to reconnect the pilot to the aircraft, "artificial-feel" systems were added to the fly-by-wire concept. The *artificial-feel system* feeds back to the side stick controller some of the forces the control surfaces experience. A system called a "stick shaker" was similar to the artificial-feel system and was used on aircraft for many years before fly by wire. This device imparted a vibration to the control yoke of large aircraft that had assist systems. The shaker would mimic the feel that an aircraft with simple, direct-connected control surfaces would experience when the air around the control surfaces became turbulent, which is a predictor of a stall.

The Lockheed F-16, whose maneuverability is legendary, was the first military aircraft to get a full fly-by-wire control system. The U.S. Air Force precision-flight team, the Thunderbirds, has flown F-16s as its aircraft of choice. Conversion of air-transport aircraft to fly by wire was slower to gain acceptance. To implement fly by wire implies that the complete control of an aircraft was put into the hands of a computer. This concept is called *full authority* and is discussed in a previous chapter relative to engine controls.

Air-transport aircraft are inherently very stable and do not need stability augmentation. However, other advantages of fly by wire could be realized. First, virtually all large aircraft have flight-control systems where, for all practical purposes, the aircraft is being flown by a computer. Fly by wire was a natural extension of the flight-control system. Second, the mechanical components of the conventional flight-control system were heavy and expensive not only to install but also to maintain and were a source of failures. In fact, as the reliability of electronic systems steadily increased, the probability of a mechanical failure in a flight-control system was greater than that of an electronic failure. Finally, fly-by-wire systems could be programmed with limits that would prohibit or at least lessen maneuvers that would possibly damage an aircraft.

Before the F-16, a number of other military aircraft, including the North American RA-5C (1958) and the Caravelle (1957), had fly-by-wire systems. The dates shown are the year of introduction of the first flying model of the aircraft, which is not necessarily the date the fly-by-wire system was installed. More important, the early fly-by-wire systems were not full authority. These early introductions either had complete mechanical backup systems or the fly-by-wire system was for controls that are not "flight critical," such as speed brakes, spoilers, flaps, and so on.

The first air-transport aircraft to have a fly-by-wire system installed was the Concorde. Its system was designed by Aerospatiale in 1969 and was completely analog. Having no history of performance of a fly-by-wire system, the Concorde was also equipped with a fully mechanical backup system.

The first commercial use of digital, full-authority, fly by wire was on the Airbus A300 in 1972. In these early fly-by-wire installations, the conventional yoke and rudder pedals were used. A sudden and drastic change in the functionality of a cockpit was not desired. Some pilots were already concerned that they were communicating their desires to a computer rather than directly to the aircraft. To use a side stick controller might have been "too much, too soon." Also, the pilot's and copilot's controls were mechanically locked together, and an artificial-feel system was provided. In later aircraft in which side stick controllers were employed, there were no mechanical connections between pilot and copilot.

Fly by Light

The fly-by-wire concept involves generating a command in the cockpit, turning that command into a signal, and then transmitting the signal to a computer, which decides what to do. After the computer has made its calculations, it sends signals to actuators that cause the control surfaces to move. Eventually the flight path of the aircraft is adjusted according to the command. Communication between the flight deck, the computer, and then the control surfaces is critical to the safe operation of the system. There is no mechanical backup. Once the communication has been cut, there is very limited, if any, flight control.

Damage to the wire would be a very serious situation. This is primarily important in military aircraft, where the communications are triply redundant. The three separate cables are routed through different parts of the aircraft and protected by armoring. Wire damage on civilian aircraft is more likely accidental than deliberate.

Another vulnerability of wired communications is interference from external radio waves. To reduce this effect, communications cables, which are usually twisted pairs, are shielded. This is common for computer networks in offices, where computers are interconnected with shielded twisted pairs, or STP. Since all aircraft have several types of radio transmitters on board, such as communications, transponders, and weather radar, the onboard cables must be shielded to protect the communications system from the aircraft's high-intensity radio waves from its own transmitters.

Strong radio waves or electric fields can be produced by nearby transmitters, such as high-powered broadcast stations, and nearby radar antennas. There is one source of very high electric fields to which all aircraft are vulnerable, and that is lightning strike. Excluding the electronics, aircraft are not seriously damaged by lightning strikes. Most aircraft that are flown regularly, such as military and air-transport aircraft, have been struck by at least one lightning bolt. It has been estimated that on average, every large passenger jet is struck once or twice a year. Other than scaring the passengers to death, no real damage is done. If any damage does occur, it is usually to the plane's avionics.

However, lightning strikes cause the most damage to systems that have long interconnecting wires because the wires conduct the energy to the system where the damage is done. Even if the energy that is contained on the interconnecting wires is not sufficient to cause significant damage to the system, one thing is for sure: A good lightning strike is going to disrupt the communications on the wire. If the signals on the wire are fly-by-wire commands, the disruption can be serious. The light signals in a fiber-optics communications system are immune to the electrical disturbances caused by lightning. Using fiber optics rather than wires for critical communications is called *fly by light*.

The Boeing 777: The Electronic Airplane

In 1986 the Boeing Commercial Airplane Group tried to interest airlines in a longer-range B-767. The 767 was at that time a well-received, economical, twin-engine aircraft. Boeing customers' experience with the 767 was that the cabin size was right, but the range was too short. The wing design did not permit the high altitude required for long-haul passenger service. A redesigned wing would increase the range to seven thousand miles, but for long-haul service, the number of seats in the 767 was insufficient. A longer-range plane would require a stretched version of the fuselage, which would increase drag and then destroy the aircraft's efficiency. The 767 was not designed for long haul, so a new design was required.

The proposed B-777 would be the latest in a long, successful line of Boeing jet air-transport aircraft. The first of the line was the venerable 707, which went into service with Pan Am in 1958. The 707 was a four-engine jet for transoceanic and domestic service. The B-727 followed, which was a tri-jet regional aircraft. After the 727, the very successful 737—a small airliner— was introduced. The 737 is still in production at the time of this writing. The 737 has been stretched, and its range has been extended; as a result, it now hardly resembles the original that first rolled off the production line. The long-range 747, also still in production, followed and became one of the most economical, long-distance, high-capacity aircraft ever. Then came the 757, which replaced the 727, and the wide-bodied 767.

The new 777 was conceived of as a twin-jet aircraft capable of trans-oceanic flight. Twin jets are considerably less expensive to manufacture and maintain. There is a serious problem with using twins on long overwater flights, however. International standards call for three or more engines to be used for such flights. Aircraft commonly used for overwater operations in-clude the four-jet B-747, the Airbus A-340, the three-jet Lockheed L-1011, and the MacDonnell Douglas DC-10 and MD-11. Twin-engine aircraft can be used for long-distance, overwater flights if they obtain special waivers called "extended-range, twin-engine operations," or ETOPS (pronounced e-tops). ETOPS requires a demonstration of very high engine reliability and the abil-ity to fly for three hours on a single engine.

The design of the B-777 marked the beginning of a unique concept called "customer as partner" in aircraft design. This program involved cus-tomers at the early phases of the design to provide them with the configuration they desired. Also, the customers would share in the product-performance data once the aircraft were delivered in order to close the loop on quality and reliability analysis. The B-777 was to be the most modern aircraft of the Boeing fleet and possibly the most modern aircraft in the world. One of the goals of the partners' program was to deliver an aircraft that was "service ready." This simply meant the customer would not be asked to help "iron out the bugs," but the aircraft would very quickly achieve product maturity.

The B-777 was to be a "paperless" design. All of the drawings would be done with computer-aided drafting, CAD. The program used was the three-dimensional one developed by Dassault Aviation and IBM called CATIA. This system features a rotatable 3D view that achieves a perfect fit of all of the aircraft components. It allows simulated operation of controls and provides access panels to check for clearances and provide ease of operation. Without the use of CATIA, wooden mock-ups were made for these purposes. The 3D views could be used to make production drawings, parts lists, test procedures and so on. Therefore, the B-777 not only was a paperless design but also featured paperless production, testing, and maintenance.

The B-777 design was to be totally new and could include the latest technology. So often an aircraft design uses components from an earlier design as a way of controlling costs. When this happens, old technology must be used to interface the newer parts with the older ones. Since no components from earlier designs were used, there was no reason to maintain any older technology that did not provide the best performance or economy. This totally new design allowed the B-777 to be the first Boeing airplane to use fly-by-wire technology.

The use of fly by wire saved weight, improved the aircraft handling, increased reliability, and reduced maintenance costs. A conventional cockpit was used with mechanically connected captain and first-officer yokes. Boeing's philosophy in cockpit design is to make similar layouts on similar aircraft types. In the case of the B-777, the layout is similar to that of another long-haul Boeing airplane, the 747-400. Although the B-747-400 is a conventional, assisted, mechanical-controlled aircraft, that plane and the B-777 fly virtually the same.

The B-777 fly-by-wire system employs *envelope protection.* This feature of the artificial-feel system provides increasingly greater force when the aircraft is pushed to its limits. A similar system is called *envelope limiting.* In this latter example, the flight-control system will not permit maneuvers beyond the limits of the aircraft. In envelope protection the pilot can cause maneuvers beyond the limits of the aircraft but is dissuaded by increasing reverse force.

The B-777's fly-by-wire system was evaluated on a modified Boeing 757. The test aircraft was fitted with the fly-by-wire system and connected to the first-officer's controls. The conventional, mechanical-control system remained connected to the captain's controls. Boeing as well as potential customers and the FAA flew this aircraft extensively. When the first B-777 was flown, the performance of the fly-by-wire system was exactly what everyone had expected.

Local Area Networks Abound

The B-777 also makes extensive use of a *local area network,* or LAN, for avionics as well as for the in-flight entertainment system. Because the 777 comprises sixty-six distinct systems, communications between systems and the systems components present a huge task. Digital communications on aircraft have been around for a very long time. The major function of airborne digital-communications systems is to transmit a number of data signals on a single wire. This process is called *multiplexing.* If every wired function on an aircraft required its own dedicated wire, many aircraft would suffer a serious weight penalty from all of these required copper wires.

Consider, for example, a large passenger aircraft. Assume that each seat has an attendant-call button and that pressing the button alerts the cabin attendants and identifies the seat. This could be done by using one switch per seat with a wire connected to an annunciator panel. If this hypothetical aircraft had four hundred seats and the average distance to the seat was 50 feet, more than 20,000 feet of wire would be required to implement this scheme. If each seat had not only a call button but also ten channels of audio entertainment, a reading-light switch, a telephone, and a video-entertainment system, it is not hard to understand how many more thousands of feet of wire would be required. In a digital network, there is one bus. Each user is connected to the bus via a terminal. Each user extracts needed data (called the *downstream data*) from the bus and transmits data (called *upstream data*) on to the bus.

Most of the standardized networks for use in aircraft were not suitable for a large number of users or for high-speed data. The goal of the B-777 was to provide a sophisticated, in-flight entertainment system, IFE, which would provide a choice of movies, television channels, high-quality music, telephone, and internet connections. Most B-777s have a highly sophisticated IFE system. The exact configuration of the airplane is up to the customer, but the ability to accept a state-of-the-art entertainment system is designed into the aircraft. The typical system features individual flat LCDs at each seat. Therefore, passengers can select from a menu of entertainment choices, including pay per view. In the highly competitive world of air travel, airlines welcome the additional, high-profit-margin revenue.

Standardized data buses have been developed for use in fixed systems such as in an office complex. Rather than invent a wholly different bus, these existing buses were adapted for aircraft use. This also opens up the possibility of using commercial, nonaircraft hardware for the IFE. Because each seat has an electronic package, the B-777 IFE system is the largest electronic system on the aircraft.

Improving Maintenance Electronically

Because the B-777 was to be the most reliable aircraft flying, it had to be in order to be ETOPS-ready when delivered. A system called the *airplane integrated-management system,* or AIMS, was developed to process the wealth of available information and generate the pertinent displays for the flight crew. The AIMS system is very complex and employs more than 2.6 million lines of computer code. The B-777 has over 4 million lines of code and 1288 embedded processors.

The AIMS system, developed by Honeywell, is an example of *integrated, modular avionics,* or IMA. Most avionics systems of the late-twentieth century consist of a collection of "black boxes." Aircraft, however, were constructed with an ever-increasing number of electronic systems. Racks of equipment were installed in the typical air-transport aircraft.

With the use of sophisticated integrated circuits, the size of equipment diminished to a point where the weight of the connectors, the interconnecting wires, the mechanical tie-downs, and the housings is more than that of the electronics. The technique for reducing weight and volume is to eliminate these very items. If ten electronic systems shared a common cabinet, there would be nine fewer cabinets and tie-downs. Although there must still be interconnections, the units are much closer to each other, so the interconnecting wires can be much shorter. Because the interconnections are within a common cabinet, the connectors can be simpler and do not have to be protected from the environment. When avionics systems are contained in individual black boxes, each box must contain most of the subsystems that allow the unit to function. This implies that many black boxes have their own power supply and computer. When systems share a common cabinet, they can share the power supply and computer, for example.

Because of the large-scale integration of electronics, entire systems can be implemented on a small, plug-in circuit card. The AIMS system consists of a "cabinet" into which a number of plug-in cards are inserted. The plug-in cards can exchange data very rapidly through the "backplane" of the cabinet. This arrangement is very similar to that of the modern personal computer, or PC. If PCs were constructed in the fashion of avionics in the black-box style, there would be one cabinet with a processor, another with a hard disk, another with a CD drive, another with a modem, and so on. Each cabinet would have its own power supply and power cord that must be plugged in the AC line. The result would be a large number of interconnecting cables and power cords and a large stack of cabinets. In the modern PC, these subsystems are inserted into connectors on a motherboard, which contains the processor.

There is a common power supply and a cooling fan. The modern computer recognizes the boards plugged in and automatically configures the hardware and software to operate the plug-in boards correctly. Diagnostic self-tests are available and determine the location of malfunctioning boards. The integrated modular-avionics approach provides not only a significant reduction in weight and volume but also increased reliability. This is because connectors and cables are some of the aircraft's more unreliable components.

The AIMS system goes beyond simply supplying information to the crew. It also collects maintenance data and makes them available to the ground crew. It does this in two ways. First, when the aircraft has arrived at the gate, ground crews can consult a maintenance terminal that is provided in the flight deck. Second, the data are sent to the ground via either the aircraft communication and reporting system (ACARS) or the VHF data link. As an example, engine parameters are transmitted to a ground network during the entire flight and analyzed. When the aircraft arrives at the gate, the ground crew knows what tools they need, which hatch to open, and what to fix.

Tolerating Faults

In a fly-by-wire or fly-by-light system, failures of full-authority subsystems can result in disasters. Therefore, the reliability of any full-authority system is paramount. However, when computers are involved in complex systems such as a primary flight control (PFC), the probability of failure increases. When an aircraft is flown almost continually for years, the probability of a failure in even the most reliable equipment is significant. To sum up the FAA's view on aircraft reliability, an aircraft should go from the manufacturer's production line to the salvage yard and *never* experience a fatal crash due to system failure. For nonfatal mishaps, an aircraft should go from manufacturer to salvage yard and seldom experience damage, such as a collapsed landing gear, or a hard landing due to system failure.

Fault tolerance means a system that will continue to function even though a significant component of the system has failed. For a system to be fault tolerant, it must be able to replace a failed component while continuing to operate. As an example, if a system contains three computers and one fails, the remaining computers can take up the tasks the failed computer would normally perform. In this example the three computers may be different, which is not the same as the previous example of three computers that were operated in parallel, performing the same tasks. This would require that the software for all of the systems be resident in all of the computers so that any one computer can take up the slack for another. This is not out of the question. Modern microcomputers have enormous storage capability. Some sort of detection scheme must be included to determine when a computer has failed, and this is also possible.

The computers aboard the B-777 are the first application of fault tolerance on an air-transport aircraft. A certain amount of fault tolerance has been used in aircraft for a long time. For example, almost every aircraft, even the smallest, has two navigation receivers. A number of navigation maneuvers benefit from two receivers, but if one fails, most of the navigation that was done with the two receivers can now be done with just one. The situation is a nuisance, but it is possible to navigate to the nearest airport or even to the desired destination and have the receiver fixed.

Although fault tolerance is somewhat similar to the hypothetical situation with the two navigation receivers, it is different. First, a computer decides to replace a failed component, and the replacement is completely transparent to the aircrew. Second, there is not always a noticeable reduction in performance after the fault occurs. This means that no one on the flight deck will know that a major component has failed, and the aircraft may be flown normally. The problem is that, once a failure has occurred, the system is no longer fault tolerant. Should additional components fail, the entire system will cease operation.

In the B-777 the fly-by-wire, primary flight control computers are triply redundant. The computers are monitored using a simple voting scheme. Other subsystems used by the fly-by-wire system are at least doubly redundant. Since the fly-by-wire system can operate with degraded components, the system operation is categorized into three modes: normal, secondary, and direct.

In the *normal mode,* all components of the fly-by-wire system are operating. The PFC computer controls the control surfaces from information supplied by the flight-deck controls. This mode of operation includes all of the protective features, such as the envelope control and ride enhancement.

In the *secondary mode,* the system has failed to a degree that the more sophisticated features cannot be implemented. The control surfaces are controlled from the inputs from the flight deck and processed through the PFC computers. In this mode the envelope control, as an example, no longer functions.

In the *direct mode,* the inputs from the flight deck are essentially electrically connected directly to the control surfaces. There is no significant processing of the inputs from the flight deck, and many of the features of the fly-by-wire system are lost.

In all three modes the aircraft can be flown safely. In the secondary mode the aircraft can be flown for hours without crew fatigue. The aircraft is usually flown to its destination, but the trip may not be as smooth as it could be. It is important, however, that the crew be alerted to the fact that the system is operating in the secondary mode. As previously mentioned, the envelope control system is inoperative and does not provide feedback to the pilot when the aircraft is pushed to the limits of performance. Although the aircraft should not be flown to the limits of its performance, the crew needs to be aware of the fact that they will get no warning if that situation should occur.

In the direct mode, crew fatigue increases significantly to the point where an unscheduled landing may be necessary. In redundant systems an errant system is disabled, and the remaining systems continue to function. Voting is the most common method of controlling redundant systems. For example, in the B-777's triply redundant PFC computer system, the voting circuits shut down the errant computer if the output of one of the three computers differs from the other two. In this situation the success of this system lies with the voting circuits. If errors occur in the voting circuits, the wrong computer could be shut down or a failure could go unnoticed. Also, the voting circuits could activate when there is no actual failure and shut down a system that has not failed. The success of a voting circuit is its simplicity. The voter controls very complex systems, such as computers, using very simple algorithms, such as comparison.

In addition, monitors detect failures before they occur. Many failures are neither sudden nor insidious. Consider a common bearing, which usually announces its impending failure by becoming very noisy. Voltages, currents, temperatures, noise levels, and so on can be monitored, and predictions of failures can be made. In the B-777 a central maintenance computer provides messages to mechanics relative to components of the flight-control system. This permits components to be replaced before they fail.

With fault tolerance, a debate centers around flying an aircraft with known failures. The concept of fault tolerance is that, even though a part of a system has failed, the overall system continues to operate. Ideally, the system is repaired at the next stop. No troubleshooting is required as the system knows which part has failed due to the central maintenance computer, and it is a simple matter to replace the failed module. The situation becomes somewhat more complex when the next stop does not have a maintenance facility. The debate centers on dispatching an aircraft with a known failure. However, the argument continues that the system has not failed but performs to its full specifications.

Although not universal, fault tolerance is increasing on air-transport aircraft at the time of this writing. We might expect that appropriate rules will be drafted to permit fault tolerance to achieve its goal: making aircraft more reliable. The rules for the B-777 are that an aircraft cannot dispatch with a secondary or direct mode.

The Aircraft as a System

We have discussed a number of systems used in modern aircraft, and many of these are in some way interconnected. As an example, let us consider the in-flight entertainment system. This system delivers the safety announcements and thus connects to the "cabin-access control panel" that the flight attendants use. Announcements from the flight deck are routed through the entertainment system, and thus the cockpit interphone is connected to the entertainment system. A very popular system is a moving-map display showing the aircraft's progress to its destination. This requires connection to the GPS receiver. In addition, the aircraft's altitude, air speed, and the outside temperature are often displayed. This requires an interface to the altimeter, the outside air temperature sensor, and the airspeed sensor.

The flight-management system has a huge number of interfaces that connect to virtually every navigation system in the aircraft. The number of electronic devices aboard the modern air-transport-category aircraft is huge. Consider the following list of common equipment items, most of which were discussed in earlier chapters:

VHF Omnirange	VOR
Distance Measuring Equipment	DME
Automatic Direction Finder	ADF
Global Positioning System	GPS
Instrument Landing System	ILS
Localizer	LOC
Glide Slope	GS
Marker Beacons	MB
Air-Traffic-Control Secondary Radar	ATCRBS
Mode S Transponder	Mode S XPDR
Weather Radar	WXR
Inertial Reference Unit	IRU
Radio Altimeter	RAD ALT
Aircraft Communications and Reporting System	ACARS
Very-High-Frequency Communications	VHF Comm
High-Frequency Communications	HF Comm
Very-High-Frequency Data Link	VDL
Traffic- and Collision-Avoidance System	TCAS
Automatic Dependent Surveillance Broadcast	ADS-B
Ground-Proximity Warning System	GPWS
Engine Instruments and Crew-Alerting System	EICAS
Cockpit Voice Recorder	CVR
Flight Data Recorder	FDR
Flight-Management System	FMS
Cockpit-Intercom System	Interphone
Aircraft-Control System (Fly by Wire)	ACS
Aircraft Integrated-Management System	AIMS

This list represents equipment discussed in this text, but the list does not tell the entire story. Many of these items are duplicated and even triplicated. As an example, every air-transport-category aircraft has two complete ILS systems. Aircraft equipped with three radio altimeters are very common.

The list is not complete. There following units are not discussed, but their nomenclature describes their function:

Cabin-Air Controller
Antiskid Braking-System Controller
Emergency Track-Lighting System
Full-Authority Engine Controller
Fuel-Management Computer

Where is all of this equipment installed in an aircraft? Most of it is in a stand-up room with racks of equipment on two, three, or even all four walls. Other equipment is scattered throughout the aircraft. If one were to take a tour of a B-777, including the cabin, galleys, flight deck, closets, and cargo area, one would see electronic boxes everywhere, hundreds of them. Two pieces of equipment that receive special treatment are the cockpit voice and data recorders, which are often placed at the base of the tail fin because this location is one of the least likely to be damaged in a crash.

Further Reading

Birtles, Philip. *Boeing 777.* Osceola, Wis.: Motorbooks International, 1998.

Bowman, Martin W. *Boeing 747.* Marlborough, England: Crowood Press, 2000.

Jarrett, Philip. *Modern Air Transport: Worldwide Air Transport from 1945 to the Present.* London: Putnam Aeronautical Books, 2000.

Norris, Guy, and Mark Wagner. *Boeing 777.* Osceola, Wis.: Motorbooks International, 1996.

Sabbaugh, Karl. *Twenty-First-Century Jet: The Making and Marketing of the Boeing 777.* New York: Scribner, 1996.

Upton, Jim. *Boeing 777.* North Branch, Minn.: Specialty Press, 1998.

Chapter 9

The Role of Electronics in the Second Century of Flight

To the pioneers of aviation in the early years of the twentieth century it would have seemed absurd to suggest that their endeavors would lead to astronauts walking on the surface of the moon or unmanned craft crawling along the surface of Mars. These pioneers had spent years just getting off the ground under power, only to find that they needed more years to resolve other problems like stability in flight. Navigating a thousand miles, much less a million or so, was difficult at best, and in bad weather all but impossible.

A similar situation existed with the wireless pioneers of the first decade of the twentieth century. Marconi struggled to span the Atlantic and could not be bothered to speculate whether signals might someday be sent to the moon. Visionaries like Nikola Tesla who suggested that radio signals could be sent to Mars were considered eccentric, their ideas as fictional as Jules Verne's stories about trips to the moon.

But humankind eventually went to the moon and astronauts sent back live pictures to earth as proof well before the twentieth century was over. Yet, some of the problems encountered by these early pioneers are still with us. Weather was always a problem for aviators, even though Jimmy Doolittle as early as 1929 made a major step toward all-weather flying with his demonstration of blind flight. But as any frequent flyer knows, weather still plays a big role in air travel. All-weather flying in the early part of the twentieth century involved the weather at the departure airport, the destination airport, and en route. In the modern air transport system, aircraft can travel from coast to coast in the United States in six hours—but the hub system creates problems. A trip from point A to point B now involves point C or D because the aircraft being used comes from C and D before taking passengers from A to B. Probably one of the biggest challenges of the twenty-first century is to effectively dispatch aircraft and recover from upsets due to weather.

The goals of the twenty-first century are not going to be simply to fly higher and faster. We have flown to ends of the atmosphere. We have flown faster than the speed of sound. The aviation industry experimented with supersonic transports, and the experiments were a costly failure. After more than thirty years, the only supersonic transport aircraft in regular service were retired. The Concorde never made a profit for its builders or its owners, and other supersonic transport projects were cancelled as a result. The dream of a hypersonic space plane has all but died. We may have reached the practical limits of higher and faster. Even so, passengers already spend more time getting to and from airports and waiting in airports than in the air.

The present communications system, which has always been an important part of safe and efficient flight, has also reached its limitations. The old analog voice systems of the first century of flight are being replaced with seamless digital communications systems that will deliver clearances, weather, traffic, and other important information to the flight deck.

Despite improved communications and dependable aircraft, safety remains another area of concern for the twenty-first century. Although commercial aviation is one of the safest forms of transportation, accidents still occur. Modern navigation prevents even the smallest aircraft from getting lost and running out of fuel. Modern radar has reduced the occurrence of midair collisions. TCAS has reduced the risk of midair collisions when aircraft are out of radar range. But TCAS is expensive and nearly impossible to install in small aircraft. ADS-B, which was discussed in a previous chapter, holds promise for an inexpensive and very effective collision avoidance system. Aircraft still fly into the ground in CFIT. The terrain awareness and warning system, TAWS, should have a profound effect on the number of CFIT accidents in large aircraft, but the challenge is to make TAWS as inexpensive as ADS-B so that it can be installed on small aircraft. The task of the twenty-first century is to produce a reliable, cost effective product that will be installed in sufficient numbers to make both general and commercial aviation safer.

Probably the biggest role electronics will play in the twenty-first century is in the area of security. It was mentioned in this text on more than one occasion that some of the most important electronic systems are on the ground not in the air. This trend will continue. Aircraft security will involve scanning devices used in airport screening, such as enhanced X-ray systems, substance analyzing systems, and most important, a data base for tracking known terrorism suspects.

With the introduction of the "free flight" concept, a telecommunications network is the key to the success of this system. Networks have been the backbone of air traffic control for decades, but the requirements of the twenty-first century free flight environment will require a more advanced telecommunications network.

Will we ever see a fully automatic aircraft? Probably. Will we ever see a crewless aircraft? Probably not. Completely automatic, crewless mass transportation vehicles have been with us for quite a while. Many airport shuttle trains that shuttle between concourses are completely crewless, but they move at modest speeds, are totally contained, and only travel short distances. Many subway systems are essentially automatic, but usually have at least one crew member on board. Subway trains automatically close the doors, accelerate smoothly to reach a cruising speed, and stop at the next station. The crew member is essentially on board to push the big red "STOP" button when something goes wrong.

There is a big difference between a train and an aircraft. If something should go wrong, it is possible to stop. Stopping an aircraft is more problematic. However, there are autonomous aircraft already in flight. The military UAV, unmanned airborne vehicle, is undergoing extensive development and testing. There are problems with the vehicle but they should be solved. Could a UAV be suitable for passenger transportation? In an industry where the time from first powered flight to the moon was only 66 years, the answer to the question should be obvious.

Some of the important advances in the twenty-first century will come from general aviation. One of the reasons that high tech solutions are not immediately installed in small aircraft is their expense, large physical size, and heavy weight. Weather radar first installed on large aircraft was a challenge to adapt to smaller aircraft. Even the high-end general aviation aircraft, such as corporate jets, could not install the large antenna required for radar, regardless of cost. Improved radars permitted antennas less than half the diameter of those installed in air transport aircraft to be used in smaller aircraft. In the twenty-first century aircraft will receive weather information from powerful ground based radar systems linked to aircraft. Even the smallest aircraft will be able to display complete weather information without the installation of expensive, heavy, and drag-inducing equipment.

A NASA-funded program, currently under way, the Small Aircraft Transportation System or SATS, attempts to increase the use of general aviation aircraft by removing some of the complexities of flying. The SATS program attempts to make an aircraft more like an automobile. In an automobile the driver simply starts the engine, depresses the accelerator or brake pedal, and steers the machine to follow a roadway. The roadway system is clearly marked with traffic control devices and the motorist arrives safely at the destination.

In the typical twentieth-century airplane, even starting its engine is not simple. Many engines needed a primer to aid in starting. The fuel-air mixture of the engine needed to be manually set. When descending with partial throttle, the carburetor was heated to prevent freezing. The FADEC system discussed in this text automatically controls all these functions, and starting and using an aircraft engine is now more like an automobile engine.

There are airways that aircraft can fly but doing so is a bit more complicated than simply steering an automobile. Modern electronics can make the task of flying a course as simple as steering a vehicle. The technique is called the highway in the sky or HITS.

It is unlikely that the SATS program will result in an airplane in every back yard or even many back yards. However, it is likely that the SATS technology will be used to simplify flying small aircraft and make it safer to fly all types of aircraft.

Virtually every advancement expected in the twenty-first century will be made possible by application of electronics. Just as the pioneers of powered flight would have never dreamed of space travel or the state of aviation at the close of the twentieth century, at the dawn of this century, we have only a vague idea of what the next century will bring. However, it has been a one hundred year trip for electronics to bring us to this point and the advancement in electronics technology shows no sign of slowing.

Glossary

ACARS   (a-cars) airborne communications, addressing and reporting system

ACAS   (a-kass) airborne collision avoidance system

ADI   attitude director indicator, or attitude direction indicator

ADS-B   automatic dependent surveillance broadcast mode

AGL   above ground level

AI   airborne interceptor

AIMS   (aims) airborne information management system

ARINC   (air-ink) Aeronautical Radio Inc.

ATCRBS   (at-crabs) air traffic control radar beacon system

ATM   air traffic management

BCAS   (bee-kass) beacon collision avoidance system

BINAC   (bi-nak) binary automatic computer

CAA   Civil Aviation Authority

CAD   computer aided design

CATIA   (ca-tee-a) a computer aided design program

CFIT   (see-fit) controlled flight into terrain

Clutter   a large number of radar returns usually from objects on the ground often called ground clutter

COMINT   communications intelligence

CRC   cyclic redundancy check

CRT   cathode ray tube

CVR   cockpit voice recorder

CW   continuous wave

DF   direction finding

DG   directional gyro

DME   distance measuring equipment

DOT   sometimes pronounced as dot. Department of Transportation

ECM   electronic counter measure

ECCM   electronic counter-countermeasure

EICAS   (eye-kass) electronic instrumentation and crew alerting system

ETOPS   (ee-tops) extended twin operations

EMC   electromagnetic compatibility

FAA   Federal Aviation Administration

FADEC   (fay-deck) full authority digital engine controller

FDR   flight data recorder

FIDO   fog, intense, dispersion of

FLIR   (fleer) forward looking infrared

FMS   flight management system

FOC   full operational capacity

Fruit   friendly replies unsynchronized in time, a form of interference in radar systems

GA   general aviation

Garble   interference in secondary radar systems due to two replies arriving simultaneously

GCA   ground controlled approach

GHz   gigahertz, 1×10^9 Hz

GOES   (goes) geostationary orbiting environmental satellite

GPS   Global Positioning System

GPWS   sometimes pronounced gee-paws, Ground proximity warning system

Grass   a large number of radar returns usually on a green monochrome radar screen having the appearance of grass

HF   high frequencies

HF DF   (huff-duff) high frequency direction finding

HSI   Horizontal situation indicator

HUD   (hud) head up display

Hz   Hertz, a measurement of frequency

IC   integrated circuit

ICAO   (eye-kay-oh) International Civil Aviation Organization

IFE   in-flight entertainment

IFF   identification, friend or foe

IFR   instrument flight rules

ILS   instrument landing system

IMC   instrument meteorological conditions

JETDS   joint electronics type designation system

JSTARS   (jay-stars) joint-surveillance target-attack radar system

kHz   one thousand Hertz

LAN   (lan) local area network

LCD   liquid crystal display

LEO   (lee-oh) low earth orbiting (satellite)

LOP   line of position

LORAN   (lor-an) long range navigation

MHz   Megahertz, one million Hertz

NVG   night vision goggles

PAR   precision approach radar

PFC   primary flight control

PHOTINT   photographic intelligence

PPS   precision position service

Rabbit tracks   regularly spaced "blips" in a radar system due to a malfunctioning radar

RDF   radio direction finding, the early British name for radar

RNAV   (r-nav) random navigation, area navigation

RTCA   Radio Technical Commission for Aeronautics

SA   selective availability

SAR   synthetic aperture radar

SARSAT   (sar sat) search and rescue satellite

SIGINT   signals intelligence

SPS   standard position service

Squitter   random signals in radar systems

TACAN   (tack-an) tactical air navigation

TCAS   (tee-kass) traffic alert and collision avoidance system

TRF   tuned radio frequency, an early form of radio receiver

TSO   technical standard order

VFR   visual flight rules

VHF   very high frequencies, 30 to 300 MHz

VOR   VHF omni range

UHF   ultra high frequencies, 300 to 3000 MHz, 30 to 300 MHz before WWII

Index

A-300, 164

A-340, 166

above ground level (AGL): relative to ground proximity warning systems, 146; required for landing, 44

ACARS, 170

ACAS, 141–42

acorn tube, 72, 86

ADI: defined, 150; as part of a simulator, 155; replaced by CRT, 152

ADS-B: defined, 157; inclusion on modern aircraft, 173; part of future air travel, 176

aerial screw, 3

Aeronautical Radio, Inc. (ARINC): assignment of channels, 98; early operations, 53; pushing for use of VHF for airborne communications, 65

aeronautical telecommunications network, 159

aeronautics branch, Department of Commerce, 17

AI, 69

AIMS, 169–70, 173

airborne collision avoidance system. See ACAS

airborne interceptor. See AI

Air Commerce Act, 17

aircraft communications and reporting system. See ACARS

Aircraft Owners and Pilots Association. See AOPA

Aircraft Radio Corporation: involvement in airborne radio communications, 58-61; involvement in AN/APR, 4, 86

Aircraft Radio Division of Radio Frequency Laboratories, 45

airmail, 49

airplane integrated management system. See AIMS

air to surface vessel. See ASV

air traffic control: considerations with collision avoidance systems, 145; first examples of, 64; improvements after WWII, 98; and need for collision avoidance system, 140–41; redefin-
ing, 157; role of post WWII navigation and communications, 91

air traffic control beacon system, xii

air traffic management. See ATM

Air Transport Association. See ATA

airways: marking with radio, 36, 55; marking with light beacons, 40

Alford loop, 55

altimeter, radar, 146

altitude, above mean sea level. See MSL

altitude, pressure, 146

amateur aviators, influence after WW I, 27

amateur radio operators: experiments with broadcasting, 30; influence during early days of radio, 15; restricted to short waves, 25; use of amateur bands for VHF navigation and communications, 90

American Expeditionary Force (AEF), 22

American Radio Relay League, 26

American Telephone and Telegraph Company, 45

AN/APR-4, local oscillator radiation, 86, 160

AN range: after WW II, 90–92; as backbone, 54; early development, 38–42, development during Doolittle blind flight, 45; as part of the Link trainer, 154; providing course information, 81–82; role in a viable airline industry, 62–64; transmission line (TL) version, 55; in WWII European theater, 67–68

antenna, trailing wire, 33

AN/TPS-10, 97

AOPA, 99

aperture, (radar), 111

Armstrong, Edwin Howard: Alpine N.J. laboratories, 46; comments on number of airborne radio equipment controls, 61; during WW I, 23–24, 35; member of Radio Club of America, 16; superheterodyne, 29–31, 33; transatlantic tests, 26; in U.S. Army Signal Corps, 22

Arnold, General H. "Hap," 106–107

artificial feel system, 163

ART-13, 61

ASV, 69

ATA, 99

ATCRBS, xii: basis for BCAS, 142–43; facing obsolescence, 138–40; inclusion on modern aircraft, 173; not suitable for TCAS, 145

ATM, 158

ATN, 159

attitude direction indicator. *See* ADI

audion, 24

automatic dependent surveillance, broadcast. *See* ADS-B

automatic flight control, 127

Aviation magazine, comments relative to depression, 50

avionics: xi, defined, 96

B-1, 120

B-17, 95, 106; on loan to MIT, 97

B-24, 95, 106

B-29, 95, 108

B-36, 108

B-46, 108

B-50, 108

B-247 (Boeing 247), 62–63

B-307, 95

B-367-80, 96, 103

B-707, 166

B-727, 166

B-737, 166

B-747, 166

B-767, 166

B-777, 166–74

batteries for radio receivers, 31, 33

BCAS, 142–43

beacon collision avoidance system. *See* BCAS

Beech, Walter, 17

Bell, Alexander Graham, 5

Bell (Telephone) Laboratories: invention of the transistor, 104; involvement with magnetron, 74; and popularity of portable electronics, 159

Bell Telephone Company, 60

Bendix, airborne weather radar, 97

BINAC, 118

black box, 169. *See* CVR, FDR

Bleriot, Louis, 13, 17

blind flight, 42–43

Boonton Township, (N.J.): role in Doolittle blind flight, 45; Aircraft Radio Company, 60

broadcasting, influence of, 25

bug, (flying bomb), 105

Bureau of Air Commerce, involvement in formation of RTCA, 63-64

Bush, Vannevar, 74, 119

Busignies, Henri, 109

buzz bomb, 76

C-54, 95

CAA: development of a landing system, 93; introduction of technical standard orders, 102; specifying communications protocols, 99

CAD, 167

Caldwell, Louis, 52

Caravelle, 164

categories, ILS, (CAT I, CAT II, CAT III), 94; applied to GPS, 127; compared to satellite navigation, 123; and ground proximity warning systems, 146; using head up display, 152

cathode ray tube. *See* CRT

CATIA, 167

central maintenance computer, 172

central offices, use in telephone system, 5

Cessna, ix

Cessna, Clyde, 17

Cessna Aircraft Company, 60; airframe for a flying bomb, 106; purchase of Aircraft Radio, 116

CFIT: defined, 146; reduction of, 176

chaff, 75

chain home, 68, 71

channel, (radio), 15

Churchill, Winston: comments on the price of the secrets of the magnetron, 73; requested radar sets from Eisenhower, 77

Civil Aviation Authority. *See* CAA

Cleveland Municipal Airport: instrument flight from, 51, 53; one of first airports under traffic control, 64

clutter: xi, 75

cockpit voice recorder. *See* CVR

Collins, A. Frederick, 20

Collins, Arthur, 60

Collins Radio Company: design of early collision avoidance system, 142; introduction of the flight director, 150; production of communications equipment, 60–61

collisions, midair, 140

COMINT, 109

command set, 60

communications, 4, 58

compass, part of the Link trainer, 154
Concorde, 164, 175
cone of confusion, (VOR), 56
cone of silence, AN range, 41, 56
conflict probe, 158
Console, 83
Constellation, Lockheed, 95
continuous wave. See CW
controlled flight into terrain. See CFIT
Convair 240, 95
Coolidge, President Calvin, 17
CRC, 139
crossband, 99
CRT, 151–52
Curtiss, Glenn, 13, 17
CVR, 148, 173
CW, continuous, 24
cyclic redundancy check. See CRC

da Vinci, Leonardo, 3
Dash 80, 96
DC-2, 62–63
DC-3, 59, 95
DC-4, 95
DC-10, 166
Decca, 83, 90
De Forest, Lee: invention of the audion, 24; using
    magnetic fields to circumvent patents, 72
Dehaviland Comet, 96
Department of Commerce: begins assigning radio
    frequencies, 29; established aeronautics branch,
    16; involvement with formation of RTCA, 63
Department of Transportation, xi
Deperdussin, breaking the 100 mile per hour bar-
    rier, 14
depression, radio and air travel during, 51
DG, part of the Link trainer, 154–55
Dingley System, fog dispersal, 43
directional gyro. See DG
dirty birds, (U2), 110
distance measuring equipment. See DME
DME, 95; inclusion on modern aircraft, 173; as a
    part of random navigation, 135; two-way rang-
    ing system, 124
doodlebug, 76–77, 105
Doolittle, James "Jimmy," 46, 50, 56, 84, 92, 175
Dornier, DoX, "flying boat," 49
DOT, xi
Douglas Sleeper Transport, 62
Dragon Lady, 109

Dubilier, William, 20
Dubilier Condenser Company, 20

earth inductor compass, 46
ECCM, 71, 76
Eclipse Aviation, (Bendix Radio), 93
ECM: defined, 71; applied to radar, 75–76
EGPWS, 148
EICAS, 152, 173
Einstein, Albert, 12
Eisenhower, Dwight: x, 77, 119
electromagnetic compatibility. See EMC
electronic counter-countermeasure. See ECCM
electronic countermeasure. See ECM
electronic warfare (EW), 24, 71
eliminate range zeros, 142
ELT, 149
EMC, 160
emergency locator transmitter. See ELT
engine instrumentation and crew alerting system.
    See EICAS
enhanced ground proximity warning system. See
    EGPWS
envelope limiting, 167
envelope protection, 167
ETOPS, 166–69
extended-range twin operations. See ETOPS

F-16, 165
FAA, xi; concern with interference from portable
    electronics, 160, 179; concerns with software,
    133; encouraging ground proximity warning
    systems, 146–47; mandating collision avoid-
    ance equipment, 157; mandating the installa-
    tion of airborne equipment, 142–43; rescind-
    ing the mode-S mandates, 140
FADEC, 138, 177
Farnsworth, Philo, 51
fault tolerance, 135, 170
FCC. See Federal Communications Commission
FDR, 148, 173
Federal Aviation Administration, xi
Federal Communications Commission: assign-
    ment of VHF airborne radio frequencies, 90;
    and emissions of electromagnetic energy, 160;
    involvement with formation of RTCA, 64;
    successor to Federal Radio Commission, 16
Federal Radio Commission: creation in 1927, 16;
    involvement with formation of Aeronautical
    Radio Inc., 52–53; regulates radio users, 29;

Federal Radio Commission (*continued*)
    replaced by FCC, 64; restriction of radio
    amateurs to wavelengths shorter than 200 me-
    ters, 25
feedback, 23
fiber optics, 166
fire control computer, 77
fix, (navigation), 81
flight critical, 164
flight data recorder. *See* FDR
flight director, 150
flight management systems. *See* FMS
FLIR, 113
fly by light, 165, 170
fly by wire, 136, 161–62; and fault tolerance,
    170–71; use in selected aircraft, 164, 167
FMS, 128, 173
"fog, intense, dispersal of," FIDO, 43
fog, problems with at airports, 42
forward looking infrared. *See* FLIR
"four course range." *See* AN range
free flight, 158, 176
frequency, 14, 29
frequency agile, 76
fruit, xi
full authority, fly by wire, 164
full authority digital engine controller. *See* FADEC

garble, xi, 139
GCA, 84, 85
gee, 81, 83
general aviation (GA): 131, 139
General Electric Company: invention of the origi-
    nal magnetron, 72; manufacturing for missles,
    117, 118
glass cockpit, 150
glide slope, 47
glitches, 131
global positioning system (GPS), 121, 123
Godley, Paul, 26
GOES, 149
"golden days of radio," 29–30, 118
GPS, 148–49; and ADS-B, 157–58; differential,
    125–26; inclusion on modern aircraft, 173; ref-
    erence locator, 127
GPWS, 146–47, 173
grass, xi
ground-controlled approach. *See* GCA
ground paint, 111
ground proximity warning system. *See* GPWS

ground waves, 54
Guggenheim, Daniel, 42–45, 47
Guggenheim Foundation, 43
gun laying radar, 72
gyro, fiber optic, 121
gyro, ring laser, 121

half duplex, 99
Hallicrafters Company, 60
Hallicrafters S-27 UHF receiver, 85–86, 109
Halligan, Bill, 60
Hammarlund Manufacturing Company, 60
"ham" radio operators, 15
Handley-Page, Frederick, 13
head up display, 152–53
Heaviside, refining Maxwell's equations, 8
Helms, J. Lynn, 143
Hertz, Heinrich, 6, 9, 14, 29
hertzian waves, 6; Marconi's successes using, 9–10;
    National Bureau of Standards investigating use
    for navigation, 29
HF DF, xii, 109
"highway in the sky," 177
HIS, 150, 152
homing, 21
Hoover, Herbert, Jr., 53
horizontal situation indicator. *See* HSI
huff duff, xii, 109
Hull, Dr. Lewis, 46–47
hyperbolic navigation, 81

IBM, Federal systems, 118
IBM, PC, 134
ICAO: formation of, 91; standardization of navi-
    gation aids, 93
identification friend or foe. *See* IFF
IFE, 168
IFF, 80; as the precursor to ATCRBS, 138; used in
    air traffic control, 96
IFR, 132
ILS, 56, 128
IMA, 169
IMC, 57
inertial navigation system, 120, 121
inertial navigation system, ship's. *See* SINS
in-flight entertainment system. *See* IFE
Institute of Radio Engineers (IRE): formation of,
    30; technical journals and early blind flight, 45
instrument flight, 39, 52
instrument flight rules. *See* IFR

instrument landing system. *See* ILS
instrument meteorological conditions. *See* IMC
integrated circuit, 104
integrated modular avionics. *See* IMA
International Civil Aviation Organization. *See* ICAO
interphone, 59, 173
interrogation, 80
intruder aircraft (TCAS), 145
ionosphere, GPS signals, 125

James Millen Company, 60
jamming, as electronic countermeasure to radar, 71
Jett, E. K., 64
Johnson, Clarence "Kelly," 110
joint electronics type designation system (JETDS), 60
Joint-Surveillance, Target-Attack, Radar System. *See* JSTARS
JSTARS, 112
Junior Wireless Club, Limited, 15

KDKA, first broadcast station, 25
Kelsey, Benjamin, 46–47
Kettering, Charles, 105–106
Kill Devil Hill, 9
Kitty Hawk, North Carolina, 9, 11
knickebein, 81
Kollsman, Paul, 44
Kollsman Instrument Company, 44

L-749, 95
LAAS, 127
LAN, 168
landing system: activities after WWII, 91; early development of, 45
LCD, 152–53; use in B-777, 168
Lesh, L. J., article, "The Development of Radio Sets for Aeroplanes," 20
lightning strikes, 165
Lilienthal, Otto, 7
Lindbergh, Charles, xi; earth inductor compass, use of, 93; full-flight laboratory, 42
line of position. *See* LOP
Link trainer, 154
liquid crystal display. *See* LCD
Litton, 118
local area augmentation system. *See* LAAS
local area network. *See* LAN
localizer: demonstration of by National Bureau of Standards, 47; having voice capability, 98

Lockheed, 95, 110
Loomis, Alfred, 81
LOP: defined, 82; in a VOR system, 94; in satellite navigation, 122
LORAN: allowing random courses, 158; compared with other navigation systems, 122; development of, 81–83; differential LORAN, 126
LORAN-C, 83; compared to other navigation systems, 122; as a part of random navigation, 135
LORAN chain, 82
"Lucky Lindy," ix

magnetron, 72–74
Magnevox, 118
Marconi: experiments with Hertzian waves, 6–7, 9–10; radio for point to point communications, 31
Marconi transatlantic tests, 10–12, 23
Marconi Wireless Company, 10
markers, microwave, 92
Martin, Rex, 63
Massachusetts Institute of Technology. *See* MIT
Maxwell, James Clerk, 6
Maxwell's equations, defining equations for Hertzian waves, 6–7, 10–11
microcontroller, 130
microprocessor: Advanced Micro Devices, 134; Intel 4004, 129; Intel 8008, 132; Motorola, 134; multiplexing, 129
microwave measurements, 80
microwaves, 73
MIT: design of an aircraft simulator, 117–18; development of landing system, 93; formation of Rad Lab, 75
Mitchel field, 42, 46
MITRE Corporation, 139
mode-S, 139, 173
Montgolfier, Etienne, 4
Montgolfier, Joseph, 4
Moore, Gordon, 132
Moore's law, 132
Morell, Dr. John, 141
MSL, 146
multiplexing, (LAN), 168

National Bureau of Standards: airborne radio, 32–33; involvement in early aviation projects, 45; landing system, 47, 92–93

National Defense Research Committee. *See* NDRC
National Radio Company, 60, 85
national standards, 57
Naval Research Laboratory, 109
navigation, long distance, 81
navigation message, GPS, 124
navigation technology satellites, 123
Navstar, 123
NDRC, 73–74
night effect, AN range, 54
night vision goggles. *See* NVG
Nobel Prize, awarded to Marconi, 8–9
Norden, Carl, 78
Norden bombsight, 73, 78–79
Northrup Aircraft Corporation, 118
NVG, 113–14

Oboe, 83
omni. *See* VOR
Omega, 122, 126
1BCG, 26
ornithopters, early flight, 3
OSCAR-1, 119
own aircraft, 145; shown on HSI, 150; in conflict
     probe, 158

P-38, 108
P-51, 108
PAR, 84
passive imaging, 112
patent battles, Curtiss and Wrights, 13
patent number 7777, Marconi's famous patent, 10
Philco, 118
photographic intelligence. *See* PHOTINT
PHOTINT, 110–11
pinger, 149
platform, inertial navigation system, 121
platform, tumble, 121
POI, 113
POZIT, 78
precision position service, 123
primary flight control, 170–71
primary flight control computer, 171
probability of intercept. *See* POI
project 621, 123
precipitation static, AN range, 54, 63
precision approach radar. *See* PAR
programmable gate array, 133
proximity fuze, 77; environment similar to ballis-
     tic missile, 104

pseudo code, 144
pseudorandom noise, (GPS), 125

Quimby, Harriet, early aviator, 12

rabbit tracks, xi
radar, xi
radiation, effects of on electronics, 120
Radiation Laboratory (Rad Lab): closing in 1945,
     96; development of microwave systems,
     80–81; formation of, 74–75; von Karman's
     comments on, 107
Radiation Laboratory Series, 81
radio, airborne, problems related to, 31–32
radio amateurs, ix; use of 160 meter band for
     LORAN, 82
Radio Club of America: formation in 1909,
     15; early articles on airborne radio sets, 20;
     proceedings of, 20; transatlantic short wave
     tests, 26
Radio Corporation of America (RCA): airborne
     weather radar, 97, 118; development of acorn
     tube, 72; development of omnirange system,
     56; experimental range station using short
     waves, 92; pioneer in broadcasting, 31; proxim-
     ity to Doolittle blind flight, 45; use of acorn
     tubes, 86
radio direction finding: early experiments, 21;
     distinguished from radio distance finding,
     68
radio distance finding, 68–69
Radio Frequency Laboratories: selected by
     Guggenheim for Doolittle blind flight,
     45–46; involvement in airborne radio com-
     munications, 58
Radio Technical Committee for Aeronautics. *See*
     RTCA
radiotelephone, airborne, evaluation by Arm-
     strong during WW I, 23, 35
Raiche, Bessica, early aviator, 12
random navigation. *See* RNAV
Raytheon, 118
RDF, early experiments by U.S. Army Signal
     Corps, 21, 68
Reaction Motors, 117
Rebecca/Eureka, 83
redundancy, 137
regenerative radio receiver, 23
reply, (IFF), 80
Research Laboratory of Electronics at MIT, 97

resolution advisory, (TCAS), 145

RNAV, 135

RTCA: evaluating software, 133; formation of, 63; formation of special committee for VHF communications, 99–100; formulating performance requirements, 102; interference from portable electronics, 160; mode-S system, 139

RTCA, DO-178, 133–34, 144

RTCA, DO-181, 139

RTCA, DO-185, 144

RTCA, SC-147, 143

runway visual range. *See* RVR

RVR, 94

SABE, 78

Santos-Dumont, 1906 flight, 13

SAR, 111

SAR, Doppler shift, 112

Sarnoff, David, 31

SARSAT, 149

SATS, 177

Scott, Blanche Stuart, early aviator, 12

SCR-520, 75

SCR-584, 77

SCR-615, 97

SCR-720, 75

search and rescue satellite. *See* SARSAT

selective availability, (GPS), 125

selsyn, 155

Shoran, 83

shore effect, AN range, 55

shrimp boats, 100–101, 103

SIGINT, 84–85, 87; from high altitude aircraft, 114

signal integrity monitoring, 57

signals intelligence. *See* SIGINT

signature, 87

simplex, 99

simulator, 155

SINS, 121

Skunk Works, 110

sky waves, AN range, 54; and LORAN, 83

Small Aircraft Transportation System. *See* SATS

spark transmitter, 14

Sperry, Elmer, 44

spread spectrum, (GPS), 125

Sputnik, 119

squitter, ix

SR-71, 115

stability, (aircraft), 162

stability augmentation system, 162

stabilized approach bombing equipment. *See* SABE

standardization, 98

stick shaker, 163

Stokes, W. E. D., "Weddy," 16

stores management, 121

Stratocruiser, (Boeing), 95

Stratoliner, (Boeing), 95

superheterodyne: invention by Armstrong, 24; not used in early airborne receivers, 35; problems with unwanted radiation, 86; use of without license, 30

synchro, 155

synthetic aperture radar. *See* SAR

synthetic vision, 114

systems integrators, 115

TACAN, 94, 124

tactical air navigation. *See* TACAN

tau, (collision avoidance), 145

Taussig, Charles, "The Book of Radio," 1922, 25

TAWS, 148, 176

TCAS: 143–44; inclusion on modern aircraft, 173; part of future air travel, 176; as precursor to ADS-B, 157–58; reducing aircraft separation, 146–47

technical standard orders. *See* TSO

technical standards, 101

telegraph, use in transportation systems, 4

telephone, early systems, 5

terrain awareness warning system. *See* TAWS

Tesla, Nikola, 10

Texas Instruments, 118, 159

Timation, 123

traffic advisory and collision avoidance system. *See* TCAS

transceiver, defined, 61

Transcontinental and Western Airlines (TWA), 65

transistor, 104

Transit, 122

transponder, 80

TRF, 30, 33

triangulation, 21–22

TSO, 57, 102

U2, 109–10, 112, 114, 115

U2, CIA, 110

UAV, 177

Univac, 118

University of Birmingham, 73
unmanned airborne vehicle. *See* UAV
U.S. Army Signal Corps, 21–22
U.S. Post Office: beginning of regular night air
    mail service, 38; cost of airmail service, 49;
    early air mail radio station, 32; initiating air
    mail service, 28
U.S. Senate bill S7243, 16

V1, 105
V2, 105, 115, 117
Vanguard satellite, 119
Verne, Jules, 3
VFR, 139
VHF data link, 170, 173
victor airways, 91, 122
visual flight rules. *See* VFR
Voissin, Gabriel, 13
von Braun, Werner, 117
von Karman, Theodore, 107, 119
VOR: and the concept of waypoints, 122; emerg-
    ing navigation method after WWII, 90-91;
    and flight management system, 128, 135; inclu-
sion on modern aircraft, 173; providing course
    information, 81-82; TACAN similar to, 94;
    used for conflict prediction, 158; VHF omni
    range, 56–58
VT (variable time) fuze, 78

watchdog circuit, 132
waveguides, 80
wavelength, relationship to frequency, 29
waypoint: 122, 134
Western Electric: involvement in large scale pro-
    duction, 60–62; manufacturer of airborne
    VHF communications equipment, 65; prox-
    imity to Doolittle blind flight, 45
Westinghouse Manufacturing Company: air-
    borne weather radar, 97; pioneering in broad-
    casting, 30
wide area augmentation system. *See* WAAS, 127
window, 75
Wright, Orville, 8, 12
Wright, Wilbur, 8, 12

Zworykin, Vladimir, 51